PRACTICAL DEER MANAGEMENT

PRACTICAL
DEER
MANAGEMENT

Charles Smith-Jones

Quiller

Dedicated to the memory of

Alex Jagger

1944 - 2018

A treasured friend and mentor to so many of us,
as well as an outspoken, passionate and dedicated deer aficionado
and a leading light in British Deer Society and Services Branch training

Copyright © 2019 Charles Smith-Jones

First published in the UK in 2019
by Quiller, an imprint of
Quiller Publishing Ltd

British Library Cataloguing-in-
Publication Data
A catalogue record for this book is
available from the British Library

ISBN 978-1-84689-299-8

The right of Charles Smith-Jones to
be identified as the author of this
work has been asserted in accordance
with the Copyright, Design and Patent
Act 1988

Edited by Kirsty Ennever
Designed by Guy Callaby

Front cover photograph:
George Trebinski
Back cover photograph (left):
George Trebinski

Printed in China

Quiller
An imprint of Quiller Publishing Ltd

Wykey House, Wykey,
Shrewsbury SY4 1JA
Tel: 01939 261616
Email: info@quillerbooks.com
Website: www.quillerpublishing.com

Contents

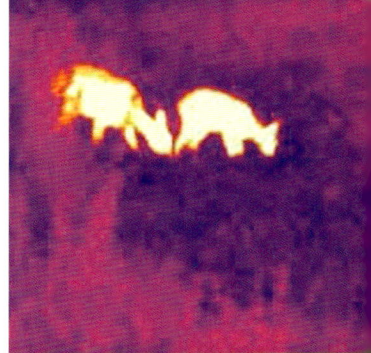

Foreword *by Ray Mears*

There are a few natural events that I look out for each year and one of my favourites is the tiny track of a new born roe deer early in May. I am not sure quite why I am so moved by such a sight, just that I am. Symbolising all of our deer species, it is a sign of continuity, of regeneration, of hope and the value of wild places. Yet I also know that the kid that made the tracks will face a difficult life. Already it will be at risk from the fox, and should its mother leave her young offspring couched in a meadow, from tractor-driven mower blades. Sadly, and increasingly, it may also become the victim of human hatred and frustration. More than once I have found lead bird shot beneath the skin of a deer where someone has lashed out with unnecessary cruelty. Yet despite these many challenges the roe deer, like all of our deer species, are survivors. In fact, our deer are an evolutionary success story, developing as they have to survive the relentless pressure of apex predators. When our ancestors removed predators to preserve our growing herds of livestock, the delicately balanced relationship between our wild deer and predators was broken. As has so often been the case, we tackled one problem only to create a new one.

Today deer are in many ways a victim of their own success, frequently considered a pest species as they impact crops, gardens, ancient woodland and contribute to road traffic accidents. Inevitably this frequently results in their indiscriminate shooting, which while providing a temporary fix can actually compound the problem.

Deer management aims to provide a win-win solution for both people and deer. As is always the case, knowledge and education are key. In this long overdue book, my old friend Charles Smith-Jones provides pragmatic, real-world advice for solving conflicts with deer. Refreshingly written in 'everyman' language, it is straightforward and easy to understand. That Charles loves deer is immediately apparent; drawing on a lifetime's experience his heartfelt belief that our deer deserve to be managed respectfully with a humane approach underpins his writing. I am pleased that he has not neglected the non-lethal control options. Not everyone wants to kill deer, and in some cases shooting may in fact be impractical, or impossible.

Deer stalkers will also find much of interest here. Over the last few decades deer management within the UK has been revolutionised by the hard work of the British Deer Society and its allied organisations. One of the most beneficial results has been the widespread acceptance of best practice education for deer stalkers, with

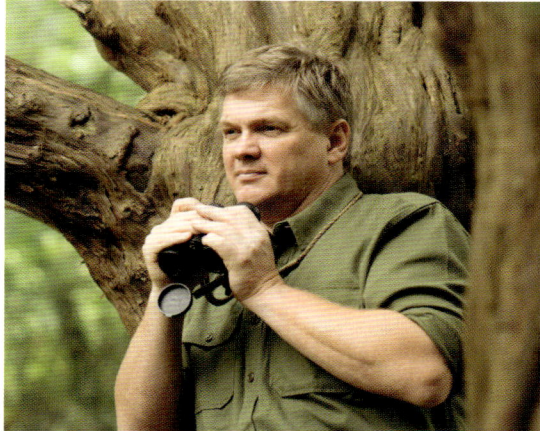

Deer Stalking Certificate Level 1 and the excellent and internationally unique DSC Level 2. Both voluntary awards, their widespread adoption reflects the willingness of British deer stalkers to demonstrate the highest professional standards of skill and competency. Despite this, there is still more work to be done, most particularly in assisting active deer stalkers to better understand the dynamics and aims of long-term deer management. Conceptually deer management is a simple process, but in practice it is anything but. We need look no further than our wild fallow deer to see this.

Fallow are highly mobile, intelligent deer that require above average stalking skills, particularly the does. Bucks are more easily culled, particularly during the rut when the testosterone coursing through their veins disables their more normal cautious alertness. These facts, combined with a very long hunting season for the bucks, result in too many antlered bucks being culled, while the does, the breeding force of the herd, may get off more lightly. Consequently, across Britain it is common to encounter large herds of fallow with many more females than males and with pitifully few males with good heads of antlers.

There are many reasons why deer managers struggle to achieve similar results nationally, but the greatest reason of all is lack of structure in their selectivity, a subject admirably addressed herein. By contrast where deer managers succeed, such as in the New Forest where the deer are expertly managed, we see modest-sized herds and very noticeably, bucks with fine sets of antlers. Now it is a simplistic assertion, but nevertheless a valid one, that the sight of good heads often indicates successful management, a sign which may also reflect a healthy environment overall.

Through this refreshing book there is the possibility for us all to find a better understanding and tolerance of all of our deer. We will never be able to stop them eating our crops or garden flowers, but we will be able to mitigate such damage confident in the knowledge that we are doing so in the best way possible, a way in which the deer may also benefit. If we succeed, we will know when we once again see fine crowns adorning the heads of our wild fallow bucks.

Ray Mears

DSC 1, DSC 2, BDS LANTRA DEER MANAGER

Preface

Whatever your involvement with the land, the chances are that deer will have an impact on you in some way. Many people have a vested interest in deer management, among them landowners, farmers, estate managers, gardeners - the list can go on and on, and their concerns might range from minimising damage to prize roses or the vegetable patch to landscape scale issues. Other considerations may include overseeing management operations, ensuring good practice, protecting agricultural crops or perhaps looking to achieve an income from venison or commercial stalking opportunities. The fact is that wherever deer are present, they will affect most rural (and indeed some urban) land-users to some extent.

While it is a potentially uncomfortable truth that the physical reduction of deer numbers might be a very necessary element of a management strategy, this could just be part of a much bigger picture and often a non-lethal option is available. Furthermore, many of the people responsible for the control of deer and minimising their impact quickly find that they are looking for a more structured approach beyond simply reaching for a rifle. It does not matter if the land concerned is only a relatively small acreage or many times that - the basic principles are very much the same. There is great satisfaction in knowing that your efforts are planned with a specific outcome in mind, whether it is to solve a problem, balance a population, reduce it or even to improve its overall quality.

Wherever deer have an impact on human interests it is a commonly held misconception, not helped by some high-profile commentators and sensational media reports, that this always tends to be detrimental. Nothing could be further from the truth. Deer need not be a liability and can often be a great asset from an environmental, and indeed an economic, point of view. Their grazing and browsing may actually hold invasive plant species in check, making way for others which provide for the specific needs of, say, butterflies and moths that select such species for their caterpillars. Likewise, the landowner can often make agricultural or forestry losses more acceptable by arranging for deer to provide payback through their amenity value, stalking revenue or venison yields. An informed and proactive approach to any perceived problem might not just neutralise it but can actually turn the deer responsible into a positive asset.

I am very aware that most of my involvement with deer has been to the south of Hadrian's Wall and while much of the emphasis of this book will inevitably be more reflective of 'lowland' deer management, many of the practices I describe or recommend will have a place anywhere that there are deer issues. I hope that the reader finds them straightforward. While it is quite possible that others may disagree with some of my thoughts, that is quite understandable as we all develop our own methods. Those I describe have worked well for me over time.

In any case, there is seldom a 'one size fits all' solution. Managing deer is not an exact science and more often than not you may find that you make a plan but are then unable to stick to it. A flexible approach is more likely to be successful: there is a need to be both proactive and reactive. At times you can forecast problems and put solutions in place to prevent them. At others, however, the intelligent manager will be able to recognise when the plans are not working and change them accordingly. Deer are adept at springing the occasional surprise!

This book is not intended as a scientific work, and it is important to recognise that deer research is constantly being updated. I have, however, included some references that might assist anyone who wishes to look more deeply into a particular area, although there may well be more recent material available as well. Many excellent sources of information exist but all too often the average reader may find them difficult to access or understand. I became very aware during my time as a college lecturer that my students had to investigate a vast amount of erudite and informed, but often confusing, resources when attempting to complete some relatively simple deer management assignment. While deer in general can constitute an immense subject in their own right, what I offer here is an everyman's guide to living and working with them.

There are numerous people and organisations that have educated, influenced and enthused me over the many enjoyable years that I have been involved with deer. Although there is not the space to name all of them here, at the forefront was Alex Jagger, to whose memory this book is dedicated with deep respect and affection. I would also like to commend the British Deer Society, which continues to do essential and sterling work on behalf of deer as our foremost deer welfare charity; and I must mention too the dedicated members of their Services Branch, also known as Defence Deer Management, with which I was associated for many years and which allowed me to be intimately involved in the purest form of deer management - namely an approach aimed at straightforward environmental balance. Finally, my time with Sparsholt College in Hampshire brought me into contact with a great many people and places that widened my perspective and, importantly, showed me yet more of the many and varied approaches to deer. Thank you all.

I do need to register a few specific 'thank yous' too. I am very fortunate to know a number of acknowledged experts in their particular fields and I would like to pick out five in particular, namely Dominic Griffith, Peter Green, John Thornley, Morris Charlton and Glyn Ingram to record my gratitude for their help, encouragement and guidance in the preparation of this book. I am also indebted to all those who have helped me to fill the gaps in my own photo library, not least George Trebinski for his stunning cover picture. Finally, very special thanks must go to Ray Mears for providing such a personal and heartfelt foreword.

What follows now is intended as a practical and straightforward handbook for anyone with an interest in overseeing or conducting deer management, who might want to limit the problems caused by deer in other ways, or who simply has an interest in deer and the issues that surround them. I hope that you find the various options and approaches helpful.

Charles Smith-Jones

1 Why Manage Deer?

In Britain, deer are our largest land mammals. Few would deny that they are a graceful and beautiful element of our fauna, and many people certainly question, often loudly, the need to control them. Why, then, should we even consider this?

The fact is that deer numbers are enjoying a resurgence at the present time. This may not be obvious to the casual observer, as even our larger species are remarkably good at going unnoticed. As a prey species they shy away from contact with any potential predator and are very capable of hiding themselves away in what might seem to be the most unlikely of places. They also tend to move at the times of day when humans are less likely to be about, often feeding throughout the night and at dawn and dusk; but as soon as human activity begins to intensify, they will probably be back in cover where they can feel safe and less liable to disturbance.

Furthermore, the British landscape is for the most part not natural. It has been shaped and controlled by human activities for centuries and if we want nature to thrive within it, it follows that active, effective and intelligent management is necessary to maintain the balance. This is a simple truth that is very easily forgotten or deliberately ignored - to the detriment of the natural world in general.

Most authorities agree that there are now more deer in the United Kingdom today than at any time in the past thousand years. Just why this is so is due to a combination of factors. Importantly, deer have no natural predators in this country; wolves were probably extinct in England by the start of the sixteenth century, and the last wolf in Scotland is reputed to have been killed in 1680. Likewise the lynx, still a formidable predator of the roe deer in Scandinavia, had disappeared by around 700 AD. More recently, a series of milder winters, an increase in the woodland cover available and a change in farming practices mean that conditions have assisted deer spread and increase. To cap it all, since the late nineteenth century three new species - the sika, muntjac and Chinese water deer - have joined the three already living wild in Britain, thanks to escapes from captivity and a series of deliberate releases. They have found our climate and conditions very much to their liking.

Most people are familiar with the popular image of the Monarch of the Glen and think only of deer in a Highland context. Yet the chances are that, wherever you might be in any part of the UK, you are never that far from a deer of some species. Few travellers on our roads or railways notice the roe in the fields, the fallow on the woodland edge or the muntjac feeding by the bramble banks that flourish on the sides of motorways or railway cuttings. Even in southern England, the larger deer species can exist in herds of over a hundred animals which emerge to raid the fields in some of our more popular beauty spots once the tourists have returned to their hotels, caravans or guest houses.

↑ *The red deer is one of the most iconic British deer species as well as being our largest land mammal.*

Start looking out for the deer, however, and learn to spot the signs of their presence, and they will soon start to become all too visible.

Increase and spread

Until the Industrial Revolution, thanks to low human populations and often draconian protection, deer were relatively common in this country. As our population grew and urbanisation took over much of the landscape, the deer were forced to draw back into what was acceptable habitat to them. The roe became almost extinct in England, to the extent that it is generally accepted that most of those in the southern half of the United Kingdom are descended largely from reintroductions from Scotland and the Continent.

Around the turn of the nineteenth century things began to change, thanks to three predominant influences. The roe was reintroduced because of a vogue for hunting with hounds. Although this fashion was not long-lived, it was sufficient to re-establish a population where it had long been absent. Then, mainly thanks to a Victorian passion for the establishment of new species from around the world, we gained three new deer - the muntjac, sika and Chinese water deer. Finally, the First World War took much of the rural workforce away to fight - and a significant proportion of them did not return. A large part of our landscape, now less intensively managed, reverted to a wilder nature which benefited the deer, and numbers began, slowly, to build up again.

The explosion in deer numbers has become most apparent in the last twenty-five years or so. A succession of deer surveys, conducted by the British Deer Society (BDS), have highlighted a dramatic increase in the range of all six deer species.[1] The surveys are based on deer distribution rather than numbers, and have mapped observations of deer species in ten-kilometre squares across mainland Britain. Comparison of data recorded in previous surveys (in 1972 and 2002) indicated an astonishing spread of deer within the country and is reproduced below:

	CWD	Fallow	Muntjac	Red	Sika	Roe
Number of 10km squares at 1972	23	249	40	576	69	778
Number of 10km squares at 2002	50	568	472	844	317	1567
Number of 10km squares at 2007	136	1025	816	1200	451	2022
Annual rate of change 1972–2002	2%	1.8%	8.2%	0.3%	5.3%	2.3%
Annual rate of change 2003–2007	22.2%	12.5%	11.6%	7.3%	7.3%	5.2%

Interestingly, the most recent 2016 BDS Deer Distribution survey[2] suggested a slight decline in the presence (if not the populations) of some species across mainland UK:

	CWD	Fallow	Muntjac	Red	Sika	Roe
Overall number of 10km squares at 2016 (mainland UK)	127	897	887	1147	428	2079

Just why this is so is unclear, but human activities may have been at least partly responsible for limiting the natural ranges of some deer species. However, even allowing for instances of misidentification or poor coverage in earlier surveys, the increase in the ranges of British deer over the past forty years or so is still significant and suggests a very rapid expansion in recent years.

Meanwhile in Northern Ireland there has been an overall continuing increase in the distribution of all species (roe and CWD are not present in NI). This may be attributable to available habitat and, in the case of muntjac, human agency is considered to be responsible for their appearance in many new locations:

	Fallow	Muntjac	Red	Sika
Overall number of 10km squares at 2011 (NI only)	60	2	89	74
Overall number of 10km squares at 2016 (NI only)	66	12	105	92

1 Etherington, T., Ewald, J. & Ward, A. (2008), *Five Years of Change*, Deer, Summer 2008

2 British Deer Society (2016), *2016 Deer Distribution Survey*, available from https://www.bds.org.uk/ (Accessed 15 December 2017)

3 Defra (2003), *Current and Future Deer Management Options*. Report on behalf of Defra European Wildlife Division

4 Harris, S. et al (1995), *A review of British mammals: population estimates and conservation status of British mammals other than cetaceans* [online], Joint Nature Conservation Committee; available from http://jncc.defra.gov.uk/pdf/pub03_areviewofbritishmammal-sall.pdf (Accessed 29 May 2015)

5 POSTnote No 325 (2009) *Wild Deer*; available from: http://researchbriefings.parliament.uk/ResearchBriefing/Summary/POST-PN-325 (Accessed 12 December 2017)

Whilst it is one thing to establish the presence of deer in a given area, it is another matter entirely to ascertain their actual numbers. The habits of deer, and the habitats that they prefer, mean than most efforts to count them can end up as a rough estimate at best. Nevertheless, a report produced for Defra, the Department for Environment, Food and Rural Affairs, consolidated the existing data in 2003 and concluded that overall numbers may have more than doubled in Britain since the mid-1970s.[3] Recognising that earlier attempts at counts may have significantly underestimated numbers, the report still came to some alarming conclusions which were largely supported by a 1995 paper written by several respected academics and published by the Joint Nature Conservation Committee[4] and 2009 figures from the Parliamentary Office of Science and Technology.[5] The following simplified table shows some of the suggested population figures:

	1970s (Defra)	1990s (Defra)	1995 (JNCC)	2009 (POST)
Red deer	190,000	360,000	316,000	>350,000
Roe deer	200,000	500,000	300,000	>800,000
Fallow deer	50,000	100,000	128,000	150,000 – 200,000
Sika	1,000	11,500	26,600	≈35,000
Muntjac	5,000	40,000	128,500	>150,000
Chinese water deer	None available	< 650	1,500	<10,000

↓ Muntjac have demonstrated an ability to live alongside humans, and are now regular inhabitants of many towns and cities.

Whilst there is some disparity between the 1990s and later figures (this is hardly surprising – deer do not lend themselves to being easily counted) there is no doubt that they point to significant increases.

Furthermore, some commentators predict even further growth and spread. It has been suggested elsewhere that the English muntjac population on its own had already reached 100,000 by 2003.[6] Another forecast said that, allowing for current rates of spread and no attempt at controlling them, almost every ten-kilometre square in Britain would contain roe by 2016, muntjac by 2022, sika by 2038 and red and fallow by 2050.[7] Only areas with truly inhospitable habitat, unacceptable to a particular species, will remain unoccupied. However, this premise should be set against the fact that the human population of the UK is itself growing rapidly; only some species of deer seem prepared to live in constant close proximity to humans, and we have yet to fully understand how this will impact on the deer themselves.

It is certainly a fact that our non-native species of deer have adapted successfully to life in this country, despite apparent differences in climate and habitat to those in their countries of origin. The spread of some,

↑ *Fallow deer have been present in Britain for so long that they are now treated as 'honorary' natives.*

such as the muntjac, has been spectacular as these deer have found and occupied an ecological niche which offers wide opportunities to spread. The sika, being rather more habitat-specific, has been somewhat more constrained but, as numbers have increased, so has their ability to occupy less attractive areas. Long considered an animal which demands acid soils and damp areas of thick cover, they are now spreading out of their southern England strongholds such as the Arne Peninsula of Dorset and the New Forest into the predominantly chalk-land surrounding areas. Even the Chinese water deer, a creature more at home in our fens and arable landscapes and considered less robust in the face of the British climate, is beginning to show signs of further expansion.

Fallow have been present in Britain for so long that they are considered naturalised, but recent archaeological evidence suggests that they were first brought to these shores by the Romans with more extensive reintroductions following in the eleventh century.[8] Perhaps, given their longer presence on this island, we would be forgiven for treating them as 'native' (fossil evidence tells us that they were once here, but died out during the last Ice Age). Certainly they exist in such numbers today, at least at local levels, that they are considered a major source of agricultural damage.

6 Munro, R. (2002), Report on the Deer Industry in Great Britain, report to Defra and the Food Standards Agency

7 Ward, A.I. (2003), Increasing ranges of British deer 1972–2002, summary of paper presented at the Mammal Society Easter Conference 2003

8 Sykes, N. J., White, J, Hayes, T. and Palmer, M. (2006), Tracking animals using strontium isotopes in teeth: the rcle of fallow deer (Dama dama) in Roman Britain, Antiquity 80

Perhaps there was a window of opportunity some years ago to contain or even eradicate the less desirable non-native deer species but, for sika and muntjac at least, that window has long since passed and our only hope is to control further growth and spread. Chinese water deer, the more innocuous of the non-natives from a commercial and environmental point of view, could yet be reduced in numbers if the will was there. There is no sign, however, of any desire for such action at present and given the Chinese water deer's fragile status in its countries of origin this may yet become a decision that is taken out of our hands.

Whilst not within the scope of this book, it is worth mentioning the wild boar, a recent arrival in our countryside thanks mainly to escapes and deliberate releases. Opinion remains divided on whether their presence should be accepted, though there is no doubt as to their capacity for large-scale damage and the potential for dramatic population growth and spread. Existing in only a few scattered locations at present, and apparently establishing themselves with great success, eradication might still yet be possible – but once more the window of opportunity is passing and will probably be lost within the next decade or so. As with the Chinese water deer, there appears to be no sign of any will to take advantage of this chance. If you are affected by the presence of wild boar, the Deer Initiative offers a series of very informative advice sheets.[9]

The impact on man

Sadly, people and deer do not always live together in harmony. As the human population of our small and increasingly overcrowded island has grown, so have deer come more into conflict with us – and we with them. It would be good to say that man and deer coexist easily, but sadly that is not the case. In times gone by deer were considered an important resource both for food and for sport. Today that is no longer the case as far as the majority of our population is concerned. Whilst deer are remarkably capable of living inconspicuously alongside populated areas, inevitable conflicts arise, especially as their numbers increase. Deer have to eat and are sadly not diplomatic in what they choose to take – especially where cultivated crops or garden plants are so easily accessible.

It is difficult to provide an accurate estimate of what deer damage costs British agriculture each year. Precise figures are difficult to come by, but in 2003 Defra estimated that agricultural damage caused by deer alone had a value of £4.4million a year.[10] The greatest damage appears to be suffered by high-value vegetable crops in the east and south west of England. Unsurprisingly, fields adjacent to woodland and other suitable cover tends to be the most badly affected. In Scotland, winter damage to root crops can have the most serious economic consequences, although it has been suggested in some quarters that winter and spring grazing of cereals might actually have beneficial effects due to greater tillering and a subsequently increased crop yield.[11]

Forestry interests are also affected. The Forestry Commission in Scotland alone spends in the region of between £6 and £7million on deer management resources, of which almost £1million is accounted for by deer fencing and fencing.[12] There is, of course, a significant return against this investment in terms of venison yield; over thirty thousand deer were culled in 2013-14 and processed through a network of thirty-seven purpose-made and fully equipped deer larders.

9 The Deer Initiative, *Feral Wild Boar in England*, available from www.wild-boar.org.uk/guide_list/ (Accessed 10 February 2017)

10 *Current and Future Deer Management Options*, Defra (2003)

11 Scott, D. and Palmer, S.C.F. (2000) *Damage by Deer to Agriculture and Forestry*, report to Deer Commission for Scotland

12 Forestry Commission for Scotland (2014), *Deer Management on the National Forest Estate*, current practice and future directions document

Some examples of deer presence and damage

1. Browsing damage. Protected by a tree tube of inadequate height, this sapling has been repeatedly eaten as new growth emerges **2.** A roe fraying stock. Note the bark damage caused by the antlers, and the characteristic scraping with forefeet at the base of the sapling **3.** Bark stripping **4.** A heavily used muntjac pathway through thick summer foliage **5.** Deer will be attracted by emerging crops and the herding species in particular can cause considerable damage. Later in the year further losses will be caused by animals bedding and rolling in standing crops **6.** A fallow browse line. The height and extent of browsing damage can help to determine the species responsible and their relative density **7.** A well-used red deer trackway or 'rack' **8.** Some deer, especially red and sika, will wallow in regular places

Meanwhile householders, many of whom complain about the loss of valuable plants or vegetables from their gardens, are quick to notice the effects of increasing deer numbers, even in some surprisingly urban environments. The 2016 Deer Distribution Survey conducted by the British Deer Society showed that muntjac alone occupy parts of every ten-kilometre square used to map their presence in the Greater London area, and although it is certainly roe and muntjac which are most willing to live in close proximity to man, others have also become bolder. In some places, such as the New Forest or parts of Sussex, a hungry fallow population is becoming more and more likely to take to raiding gardens. Today it is possible to see deer, relaxed and unconcerned, couched or feeding in suburban gardens where they have learned that they are in no significant danger from the human occupants and have become increasingly audacious – even in broad daylight.

As deer and human populations have grown, so too has the more serious issue of traffic accidents. There are an estimated sixty thousand collisions involving deer and motor vehicles every year in England alone, although there is no way of quantifying the true number. Peak times for accidents tend to be between 6pm and midnight, and then 6am and 9am, the morning rush hour when there is a larger than usual number of vehicles on the roads. Some accidents, while catastrophic for the deer concerned, may result in no more than damaged bodywork for the vehicle.

➜ *Deer/vehicle collisions have become increasingly common as deer populations and road use have increased.*

While it is easy to imagine the consequences of hitting a large deer when travelling at speed, even the sudden appearance of a smaller animal might cause a driver to swerve and lose control of their vehicle. In 2007 a major report by the National Deer-Vehicle Collision Project suggested that, between 2000 and 2004, at least 20 collisions resulted in human fatalities and a further 174 caused serious injuries.[13] In economic terms, the value of insurance claims was in the region of £13.5million for England and £3million in Scotland. The average cost of vehicle repair after a collision with a deer was around £1,710. Once again, as in so many

13 Langbein, J. (2007), *National Deer-Vehicle Collisions Project England* (2003–2005), Deer Initiative Research Report 07/1

matters concerning deer, we have no way of knowing the exact figures and doubtless today the costs in both human and economic terms will be higher still.

In England, around 40% of these collisions involve fallow deer, followed by 32% roe and 25% muntjac. In Scotland the position is different, reflecting different deer population structures. There, some 69% of road casualties are roe and 25% are red deer.

It is worth stressing that wild deer are not normally aggressive towards people, although very rarely a tragic event might occur. Such incidents have generally involved animals trapped in enclosed areas, with the injured parties blocking the only line of escape, or disturbed animals fleeing in panic during pheasant shoots. Wild deer are usually naturally inclined to avoid humans. Park and otherwise captive deer, however, can become more accustomed to the proximity to man and lose their fear of him. This has occasionally resulted in injuries, almost inevitably occurring when a park stag full of adrenaline during the rut is approached too closely or has his personal space invaded.

Nevertheless, deer occasionally feature as pets, most often as a result of having been found 'abandoned' as fawns and mistakenly taken in by well-meaning people. Usually this is a grave error which will cost the young deer its life; without specialist knowledge they can be very difficult to rear successfully and a great many die. It is always best to leave the young animal untouched and where it is, as its mother is usually not far away. If the location is revisited a few hours later she will almost invariably have returned and moved her young elsewhere.

↑ *A roe buck retreats into cover. Wild deer prefer to actively avoid humans wherever possible.*

Although a hand-reared female deer can become very tame and docile (some are occasionally encountered in 'petting zoos'), it should never be entirely trusted. Although they do not have antlers, female deer are capable of delivering a potentially serious blow with their forefeet if alarmed or put under stress. The hand-reared roebuck has a special reputation for unexpected ferocity, especially as he matures and starts to become territorial; it is no accident that roe are seldom seen in zoos and collections.

The impact on the environment

We should not neglect less commercial areas. Deer, once out of balance with the environment they inhabit, can be the cause of great damage to it. This is a particular problem with the larger herding species, especially fallow, which can quickly reduce the understorey of the woodlands they frequent through heavy browsing pressure. This will have a significant effect on the other fauna present as food sources are reduced and the woods become colder, less hospitable and more exposed. Some insects, which rely on specific plants for their larvae to feed on can disappear entirely, and ground-nesting birds may recede. In extreme cases soil erosion can result as root systems decline.

Certainly, characteristic woodland plant species, such as oxlips and bluebells, can be reduced by overgrazing by deer, but as a wider structural plant diversity is lost so some bird species such as the nightingale can be badly hit. There may also be associated declines in invertebrate abundance and diversity, along with the animals and birds which depend on them as food sources. It is also important to consider that trees themselves need to regenerate; when there are too many deer the seedlings are eaten long before they have the opportunity to develop.

➜ *A deer exclosure in a nature reserve heavily populated by deer but where numbers are not controlled. Inside, plant growth is prolific while outside overgrazing and browsing has reduced the ground cover to coarse grasses less palatable to the deer.*

Overpopulation: A Case Study #1 – Maui, Hawaii

← Axis deer

Between 1959 and 1960 nine Axis deer, native to India, were released on the Hawaiian island of Maui, both for hunting and for the sake of simply introducing a new species. Having no natural predators, by 2016 the population was estimated at around 60,000 and growing at over 20% a year. Given that the species is an aseasonal breeder (like the muntjac found in the UK, which can conceive and give birth at any time of the year), and the suggestion that the Maui Axis population is 90% female, growth rates may be considerably higher in reality unless significant control measures are put in place. Sport hunters, sadly, tend to be more interested in animals with antlers – hence the sex imbalance.

Environmentalists are now deeply concerned by increasing land erosion and the associated sedimentation which damages coral reefs. The annual economic damage caused by the deer is estimated in millions of dollars and hunters are now encouraged to pursue them without seasons or bag limits to try to contain the problem. ▶

Overpopulation: A Case Study #2 – New Zealand

In the mid-nineteenth century, the Societé Zoologique d'Acclimatation was founded in Paris, the first of many acclimatisation societies set up to encourage the introduction of non-native fauna and flora around the world. Many mistakes were made, resulting in some devastating environmental effects. The reasons for wanting to do this varied. Some European settlers in colonies around the world missed familiar animals from their home countries, while others wanted to 'improve' the local wildlife or to hunt for sport. Whatever the motivation, in this way deer made their way to many countries where they did not occur naturally. New Zealand alone gained seven deer species which still flourish there today while others (such as the moose) failed to establish successfully. The new arrivals, in a country where the only naturally occurring mammals were bats, had an overwhelming impact on the available vegetation and the native species which depended on it, resulting in the New Zealand government eventually employing full-time cullers to reduce numbers. It was not until the 1980s that deer numbers were felt to be coming under any real control. ◗

↓ Fallow are just one of seven species of non-native deer currently found in New Zealand

There has been much research exploring the relationship between deer and other wildlife. In 2011 one such report, focusing on breeding bird populations, concluded that deer might be responsible for the decline of some bird species and recognised the importance of intelligent management strategies.[14]

However, it did also take pains to note that deer may not be the only culprits in the decline of some species, and that there is much still to be learned about the implications of widespread deer browsing for wider biodiversity.

Case Studies 1 and 2 illustrate the effects of deer overpopulation on two different environments, those of Maui, Hawaii, and New Zealand.

The impact on deer

It is easy to overlook the effects of overpopulation on the deer themselves. Too many not only have a negative on man and the environment - excessive numbers can also profoundly affect the deer too. At lesser levels, these adverse effects may manifest themselves in such ways as lower body weights and a poorer general condition, with increased aggression and other changes in behaviour. At higher levels, the effects are far more serious. As densities increase so does mortality, especially among juvenile animals, and birth-rates tend to fall. This latter fact is probably linked to reduced body condition brought about by nutritional stresses, which in itself will cause an increase in winter mortality.

While the larger, herding species can tolerate a degree of crowding, up to a point, our smaller deer certainly prefer to live a more solitary existence. Too dense a population creates stresses that the deer will find hard to tolerate. Territorial

14 Newson, S. E., et al (2012), Modelling large-scale relationships between changes in woodland deer and bird populations. Journal of Applied Ecology, 49: 278–286

Overpopulation: A Case Study #3 – St Matthew Island, Alaska

St Matthew Island is a remote area measuring about 128 square miles located off the coast of Alaska in the Bering Sea. In 1944 the US Coastguard released 29 reindeer onto it as a food source and for recreational hunting by their personnel stationed there. When the coastguard station was abandoned soon afterwards, the reindeer population exploded thanks to a lack of predators, and by 1963 had reached 6,000 animals. Then the lichen that the reindeer depended on for food ran out and over the course of a single winter there was a massive die-off. When researchers reached St Matthew Island in 1966 they found that only 42 animals, all but one of them cows, were still alive. A different antler casting and regrowth cycle (reindeer are the only deer where both males and females normally grow antlers) would have given the cows a competitive advantage for the fast dwindling food resources. The single remaining bull was infertile.[15] Unable to reproduce, the reindeer population on the island had disappeared by the 1980s.

⬆ Reindeer – this one is part of a free-ranging domesticated herd in the Scottish Cairngorms.

The story of St Matthew Island is frequently held up as a classic case study of overpopulation and sustainability. Today, the only resident mammals are those that were there prior to 1944 – a species of vole and the Arctic foxes that prey upon them to maintain a natural balance. ◗

Overpopulation: A Case Study #4 – Hampshire, United Kingdom

⬆ Casualty of a 'die-off'– this roe carcass has been heavily scavenged by badgers.

The problems of overpopulation do not just apply to exotic species and overseas islands. In the early 1990s I joined a small pheasant shoot that operated over an enclosed area of some 2,000 acres, with a growing population of roe deer. Ten years previously the roe had been struggling to re-establish itself in this part of southern England but had now reached the point where the available habitat simply could not sustain the increasing numbers. The final straw came one winter when every covert we went through on a shoot day contained at least one deer carcass, some very recently dead. A combination of lack of food, cold weather and increasing land usage had led to a rapid and dramatic die-off.

It transpired that the deer-stalking rights for the area were held by two individuals who shot a few bucks during the spring and summer but left the doe population untouched throughout the winter months, allowing the unchecked breeding which had led to this situation. Later estimates were that the roe population had peaked at over 300 animals; a realistic management plan, taking habitat, land use and other local factors into account, subsequently set the holding capacity of the ground at no more than 70. ◗

tensions will arise, displacement of individuals then occurs and natural seasonal movements may change too. Once numbers reach a certain point, stress may lead to debilitation and disease, even before the food naturally available may start to run out.

It follows that where there is insufficient natural predation, it falls to man to keep deer populations in check if catastrophic consequences are to be avoided. The case studies that follow illustrate the need well enough, though in two different ways. St Matthew Island shows the outcome of no control whatsoever, while the Hampshire example illustrates the effect of insufficient and unstructured deer management.

Interactions

As a general rule, deer tend to keep themselves to themselves. Interactions tend to be between animals of the same species: others are usually ignored. Of all our deer, the roe seems most nervous and intolerant of other species, particularly muntjac which are to a degree competitors for similar foodstuffs. While roe and muntjac can occasionally be seen feeding almost side by side, it is sometimes observed that the larger roe will vacate areas where muntjac densities have swollen to levels unacceptable to them despite the size difference. Roe also seem to demonstrate intolerance of fallow at times - possibly an evolutionary strategy that assisted their survival in poorer habitats.[16]

↓ Roe and muntjac can have an uneasy relationship where densities of the latter are high.

Hybridisation is only possible between two of our UK deer species, the red and the sika, which are very closely related. It cannot take place among the other four species despite occasional claims to the contrary. Some very wild suggestions in the past, which have included red deer crossing with cows or horses, and roe deer with sheep, are now known to be genetically impossible.

While wild deer and farm livestock usually take little notice of each other, sometimes circumstances bring them into contact. Red and sika deer have been known to graze freely alongside herds of domestic cattle although there is usually little if any interaction. Roe deer are said to actively avoid sheep; this is certainly partly true, but probably has more to do with the scent of intensely farmed fields. While I have certainly observed roe to have stopped using pastures once sheep had been turned into them, and avoiding them for some weeks after the sheep have been removed, they are still sometimes seen in close proximity to each other.

Very rarely, you do find individual animals becoming more interactive with farm stock. This seems to be more common with red and fallow deer and there have been cases where an individual deer appears to have 'adopted' a herd of cattle. One recent instance involved a young fallow doe which was observed to be living with a herd of cattle for a period of several months. There could be several explanations for such behaviour: one is that the animal might have lost its mother at some point during the winter and, as a herding species, adopted the cattle as a source of security, and the cattle certainly seemed to have accepted her.

15 Klein, David R. (1968) *The Introduction, Increase and Crash of Reindeer on St Matthew Island*, The Journal of Wildlife Management, Vol. 32, No. 2

16 Ferretti, F., Storzi, A. and Lovari, S. (2011), *Contact!*, Deer, Spring 2011

Photos: Aileen Collins

↑ *This young fallow doe took up residence with a herd of cattle, even staying with them when they moved pastures, for a period of several months.*

↗ *Regular interaction between the deer and cattle was noted.*

It is also possible that the farmer might have hand-reared her. There are several known instances where red deer hinds have been hand-reared after being found in silage grass and later turned out to run with the cattle, even going in and out of the milking parlour to get a ration of concentrate feed. Some disappear to take part in the rut and then return, while others stay with the cattle all year. Others are even housed with the cattle in the winter sheds.

Another known case involved a fallow buck of about the same age, though that one eventually left to join a herd of its own kind. In a similar vein there is a record of a rescued roe kid that had been put in with a barren ewe. The ewe started lactating and successfully reared the kid before it eventually departed voluntarily.

Other stories do not always end quite so happily. A few years ago, a twelve-point stag joined a flock of sheep in Cambridgeshire during the summer and stayed on until winter. When the shepherd started feeding the sheep with sheep nuts in troughs, the stag started accidentally killing sheep as he tossed his head about to monopolise the feed in the trough. In this case there was no option but to shoot the stag which, having discovered the benefits of the sheep nuts, would probably have done the same with another flock even if darted and moved.

Whilst deer are certainly capable of carrying diseases which can transfer to livestock, instances of this appear to be rare. Although wild deer are free ranging, this is actually a benefit as it does not force them into close concentrations in restricted areas where diseases can be more easily harboured and passed on. At the time of writing (in 2019), there is not believed to have been a confirmed case of Foot and Mouth disease in wild deer in the UK, and deer are not considered to have played any real part in its transmission in the catastrophic outbreaks among farm stock during the past fifty years or so.[17]

Deer welfare

Overpopulation can have a significant impact not just on man and the environment, but also on the deer themselves. This has long been recognised, and is underlined by a major examination of wild deer populations in *The Science of Overabundance* published in the United States and focusing on white-tailed deer.[18] This takes a more practical approach and clearly identifies the relationship between population densities and overall herd health. This in many ways has a more practical bearing

17 The Deer Initiative (2009) *Best Practice Guide – Foot & Mouth*; available from www.thedeerinitiative.co.uk/uploads/guides (Accessed 20 January 2018)

18 McShea, W.J., et al (1979), *The Science of Overabundance: Deer Ecology and Population Management*, Washington, Smithsonian Books

19 Eve, J.H. (1981), *Management implications of disease*, pp 413-423 in *Diseases and Parasites of White-Tailed Deer* (W.R. Davidson, et al), Tallahassee FL, Tall Timbers Research Station Miscellaneous Publication No 7

on how we can target our management with overall deer welfare in mind. A key section, quoting an earlier paper[19] that recognises the three critical phases of populations (which, once again, can be applied to many species other than deer), is worth reproducing here:

Phase I: Virtual absence of disease. When the herd is in balance with the environment, relative population density is low to moderate; reproduction, yearling antler development, body weights, and nutritional levels are high. The abdominal parasite count (APC) is low, as is the external parasite count. Disease in the deer population is negligible.

Phase II: Acute overpopulation. This is overpopulation of recent inception and short duration. Covert disease is present but usually can be demonstrated only through laboratory diagnostic procedures. After rapid herd growth, population density is relatively high and above carrying capacity; reproduction, antler development, body weights, and nutritional levels remain relatively high; APCs are high to very high; and carrying capacity is being progressively reduced. There may be a lag period of several years after the carrying capacity is exceeded before physical deterioration can be seen in the deer.

Phase III: Chronic overpopulation. Overt disease frequently is evident through gross observations of any or all of the following: unusual losses of adults and fawns, depleted fat reserves, general build-up of different species of internal and external parasites, and gross lesions of internal organs due to various pathological conditions. Several years after carrying capacity has been exceeded, reproduction, antler development, body weights, and nutritional levels decline sharply; APCs are excessively high; and carrying capacity is greatly reduced. Population decline basically due to nutritional deficiencies in Phase III may be attributed to more visible agents that merely deliver the *coup de grâce* such as parasites, predators, and infectious diseases.

In essence, our efforts in managing deer should therefore aim for a situation which reflects Phase I. An occasional lapse into Phase II is not the end of the world as long as it is corrected quickly, but allowing Phase III to develop is very poor management indeed. Sadly it is not unheard of, as the Hampshire Case Study on page 24 illustrates all too well.

Disturbance, particularly by humans, can add to the stresses on the deer. Every time that a deer is disturbed and forced to abandon its normal daily activities, there is inevitably a calorific cost as it burns unnecessary energy. Research in

⬇ *Winter is a time when many animals, weakened by a lack of food and subjected to extremes of weather, are more likely to succumb.*

Denmark has indicated that a simple instance of roe deer disturbance by human activity, such as dog walking, demands that the deer may need to eat an extra 310 grams of food to replace the energy expended.[20] Longer term disturbance, involving extended or repeated occasions of flight, might require as much as an extra 2.7 kilograms. In calorific terms this translates as 0.2 kilocalories if the deer is slightly disturbed by a passing bicycle, 9 kcal if chased by a dog, or a massive 76.5 kcal if continually moved by a six-hour orienteering event. Of all human activities that take place in the countryside, orienteering is probably one of those that affects deer the most, with runners moving constantly away from normal human routes such as established tracks and trails. Deer can become remarkably tolerant of regular, recognisable human behaviour, but the unpredictable can be a major source of stress.

Poaching is an ever-present concern wherever deer are found and is a major problem in some areas. It can be particularly prevalent where good road access allows a quick departure from the scene. The old romantic image of the 'one for the pot' poacher feeding a hungry family is long out of date; modern poachers often operate in well organised gangs with no objective other than making large amounts of money from their activities. They care little for the damage they do to farmland infrastructure. Poaching is indiscriminate, with no consideration given to deer welfare or legal shooting seasons, and can seriously disrupt management plans by killing animals which might otherwise have been left as part of a structured herd. Furthermore, as it tends to take place largely at night, it can be extremely dangerous. The bona fide deer stalker cares very much where his bullet is likely to end up; the poacher does not. In some cases the carcase of the poached deer may even be irrelevant to those who commit the crime and left to lie where it fell, whether shot, mutilated by dogs or run down by a vehicle.

At the present time it is difficult to justify active methods of deer population control other than with appropriate firearms. Contraception may be an option for the future, as we will see later on, but is not viable at present. Hunting with hounds is now very much a thing of the past and does not fall within the scope of this book, although it should not be confused with the use of properly trained tracking dogs, which assist the stalker in following up a shot and recovering the deer.

Myths and misunderstandings

Every now and then more 'natural' methods for controlling deer numbers are proposed. The reintroduction of the wolf to the British countryside is a perennial favourite and, on the surface, an attractive green solution; proponents are also quick to point out benefits which, besides controlling deer, include eco-tourism, conservation and promoting the species itself. Successes elsewhere are frequently cited. In the United States' Yellowstone National Park, for example, it is certainly true that wolves, absent in the area since being eradicated in the 1920s, were reintroduced in 1995 and that since then burgeoning elk numbers have been reduced and there are signs that the overgrazed ecosystem is recovering. Yellowstone, though, is a huge wilderness area, about twice the size of Scotland's Cairngorms National Park and lacking any of the private ownership issues found in the latter. By the time that the wolf became extinct in England, probably around the end of the fifteenth century, the British human population was only some four

20 Jeppesen, J.L. (1984), *Human Disturbance of Roe Deer and Red Deer: Preliminary Results*, Communicationes Instituti Forestalis Fenniae 1984 No.120 pp.113-118

21 International Union for Conservation of Nature (2018), IUCN *Red List of Threatened Species*, available from http://www.iucnredlist.org (Accessed 1 September 2018)

million; today is it well over sixty-five million and rising. Wolves need a lot of space, which our overcrowded island simply cannot offer.

The same is true of the lynx, last seen in Britain around 700 AD but another creature that some would seek to reintroduce. However, this more solitary predator is too small to predate the fallow and red deer, the main issues in England and Scotland respectively, and there is concern that its ambush hunting style might not make much of an impact on the muntjac which favours thick cover. Certainly, in northern Europe it is the roe deer that constitutes a very large proportion of the lynx's diet, to the extent that, in some parts of Scandinavia, the roe population has reduced significantly – some would say catastrophically.

Charismatic species such as the wolf and lynx might be championed as potential tourist attractions, but in reality they tend to be extremely elusive; it is significant that many people who live in parts of Europe where the lynx is common have never actually seen one. Furthermore, one aspect of reintroducing apex predators such as the lynx and wolf to Britain that is frequently overlooked is the effect on livestock. Farm animals are easy prey, and it is not difficult to imagine the attraction of a slow-moving sheep compared with a fleet-footed deer in the eyes of a hungry predator. In many countries, compensation is paid by governments for livestock killed by wolves and other wild predators and farmers have to take considerable protective measures, such as the overnight penning of stock, that those in the UK are not accustomed to needing. Neither wolf nor lynx is considered endangered; the International Union for Conservation of Nature classifies the status of both as being of Least Concern at the time of writing.[21]

Photograph: New Forest Wildlife Park

Eurasian lynx.

An ethical approach

That deer numbers need to be controlled in places is undeniable, but an appropriate approach is essential as ill-considered management practices can result in imbalances and unacceptable levels of stress to the remaining animals. Simply to hope that a problem will go away is unrealistic, and sadly a modern aversion to killing what are undeniably beautiful creatures has in places resulted in overpopulation with all its negative effects on the environment, human activities and, of course, the deer themselves.

Deer do not die of old age. Tooth wear gradually reduces their ability to break down efficiently and subsequently digest foodstuffs, and the animal will eventually succumb to malnutrition.

Where culling is deemed to be necessary, effective deer management should always have deer welfare at its heart. Furthermore, within most plans there is usually a place for the recreational hunter, appropriately trained and efficient, who brings with them much needed money to finance the wider operations. Elsewhere, though, it may not even be necessary to pick up a gun.

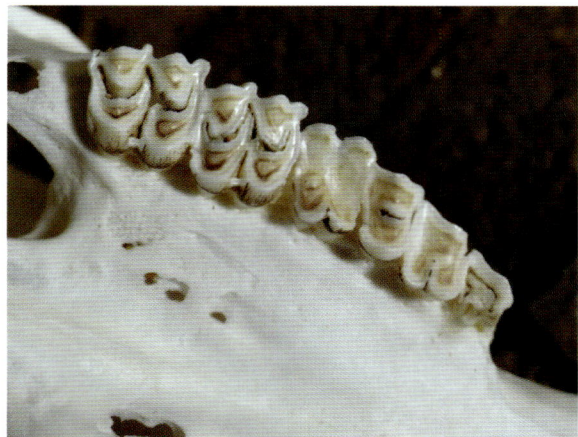

Ultimately, deer can be a valuable and sustainable resource, enhancing our landscape and more than capable of justifying their presence within it. With an enlightened approach the solutions to any problem areas are usually straightforward, and the chapters to come will suggest ways to achieve this goal. ■

2 British Deer

We have six wild living species of deer in Britain, ranging from the diminutive and secretive muntjac to the country's largest land mammal, the red deer. Each is different in many ways, some of them obvious and others rather more subtle. They certainly present a range of challenges in their management and frequently require an individual approach. A basic understanding of the natural history of each species is essential if these challenges are to be tackled successfully.

Whilst the following pages can provide only a brief overview of the species, they do offer a commentary on the particular issues associated with each of them. If you want to know more about the natural history of our deer, some recommended reading can be found at the back of this book.

← *Comparative sizes of British deer: (l to r) 6ft tall man for comparison, fallow, roe, muntjac, Chinese water deer, sika and red.*

Muntjac *Muntiacus reevesi*

Appearance

Standing about 50 centimetres high at the shoulder and weighing approximately 20 kilograms, the muntjac is roughly the size of a springer spaniel. The coat is a chestnut or gingery colour in summer but becomes thicker and greyer in winter. The tail, about 17 centimetres in length, is held vertically when the animal is alarmed to display a white underside. The legs are relatively short and the rump appears higher than the shoulder. The overall effect is a robust, pig-like appearance; uninformed observers have sometimes reported muntjac sightings as 'wild boar'!

Only the males (bucks) produce antlers, usually little more than single spikes, from the top of long pedicles, the bony extensions from the skull from which the antlers grow. These are cast in spring and regrown by the end of the summer. A black line extends up the inside of each pedicle, appearing as a V-shape on the face.

(Top) Muntjac buck.

(Bottom) A muntjac doe. Small, secretive and favouring heavy cover, this species can be one of the most difficult to manage.

➜ *British Deer Society Deer Distribution Survey 2016 results: Muntjac. Dark colours represent squares only filled in the 2016 survey, lighter colours squares reconfirmed in the 2016 survey and recorded in the 2007 and/or 2011 surveys, orange represents squares unconfirmed in 2016 but recorded in either the 2007 and/or 2011 surveys.*

1 Dick, J, Provan, J. and Reid, N. (2009), *Muntjac Knowledge Transfer: Ecology of introduced muntjac deer and appraisal of control procedures.* Report prepared by the Natural Heritage Research Partnership, *Quercus* for the Northern Ireland Environment Agency, Northern Ireland, UK.

Females (does) have a diamond- or kite-shaped dark patch on their foreheads.

Both bucks and does grow canine tusks, which are little more than vestigial on the latter. They are visible below the upper lip on the bucks, and are important to them both as weapons and for territorial marking. The skin around the neck, shoulders and rump is very thick given the size of the animal, and protects against the slashing canines of rivals.

Distribution

Introduced into Britain in the 1890s from its native China and Korea, the muntjac has adapted well to life in Britain and thrived. Although until only a few decades ago the main population was largely confined to the Midlands and East Anglia, today it has been recorded as far north as Cumbria, west into Wales and south as far as Cornwall. It has to be said that much of this spread can be attributed to human assistance and deliberate releases. Though anecdotal reports persist, there is no evidence at the time of writing (2019) that muntjac are established in Scotland, although a small and currently scattered population has recently been confirmed in Ireland.[1]

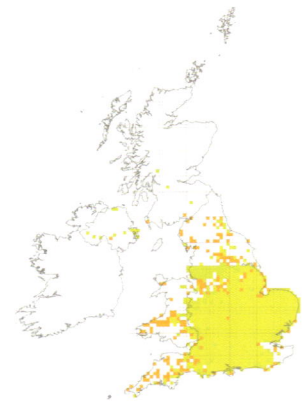

Habitat and feeding

Dense cover is preferred: thick woodlands, young conifer plantations and overgrown gardens all provide a suitable habitat in which the muntjac is comfortable. Extensive bramble beds are especially favoured, providing food, cover from view and the elements, and protection throughout the year.

The muntjac is predominantly a browser of new shoots and leaves. Having a small digestive system, it seeks out high quality fodder, from which it can extract nutrition easily; it is not attracted to coarse or woody foodstuffs or to conifer leaves. Berries, fruits, fungi and nuts are all taken according to season. Small saplings, normally too high for the animal to reach, may be borne down under the chest of a muntjac to allow it to browse the new growth on the top and then released to spring upright again. Such feeding is often mistaken for browsing damage caused by a larger deer species.

Behaviour

The muntjac is a secretive deer which is seldom seen in open country. It is increasingly the most common deer in suburbia provided that it has suitable cover to retreat to, and if unthreatened it can become remarkably tolerant of human presence. Very often the presence of muntjac in a locality can go unnoticed for considerable periods before numbers build up and animals start to be sighted.

Muntjac are restless, seldom standing still for long, and feed in short bouts before retiring to a secure place to ruminate. As befits an animal which prefers thick cover, scent-marking is an important form of communication. Males also fray saplings and suitable branches with their canine tusks to mark their territory. Both sexes make a terrier- or fox-like bark, repeated every few seconds and sometimes going on for up to an hour, giving rise to the alternative name of 'barking deer'.

They are largely solitary and, although sometimes numbers may congregate in areas containing high quality forage, these cannot be considered herds. Muntjac are more usually seen singly or in small family groups of a doe, her current fawn and occasionally her previous one. A buck may also be in attendance, but like all of our deer muntjac are polygamous; they do not pair for life and a buck will seek to mate with as many does as he can.

Breeding

Muntjac are unique amongst our deer in that they have no fixed breeding season. Does first come into season at about seven months of age and gestation lasts another seven months. Within days of giving birth to a single fawn, the doe will come into oestrus again and thus spends most of her adult life at some stage of pregnancy. Bucks, whether they have cast their antlers or not, remain fertile throughout the year and fighting between them, in which the canine tusks are used, can be fierce to establish dominance.

The fawn is weaned at about two months old, coincidentally about the age that its coat loses its infant spotted pattern. It is driven away by its mother before she gives birth again; thereafter it may be permitted to rejoin the family group. A mature doe closely pursued by a buck, especially if she appears thin, can reasonably be considered to have just come into season again, having only recently given birth.

Issues associated with muntjac

❭ High densities of muntjac are particularly damaging to the environment; they can strip the understorey of woods to the point that habitat for ground-nesting birds and small mammals is removed. In extreme cases soil erosion can result. Agricultural damage is predominantly caused by the taking of new shoots as they emerge in the fields. Generally speaking, crops become less attractive to muntjac as they mature.

❭ Of all the deer species, the muntjac seems the most readily disposed to live in close proximity to human habitation. Suburban muntjac can cause considerable damage to ornamental gardens and vegetable patches. Excluding them is difficult, and shooting may be neither a safe nor a legal option in built-up areas.

❭ Those managing game shooting often find that muntjac have a negative effect by reducing holding cover for game birds, and spoiling drives by flushing birds early by running through at critical points.

▶ Muntjac and the native roe deer may compete for the same habitat and foodstuffs. Although the roe is the larger of the two, it is more likely to be displaced.

▶ Because of their size, restless nature and secretive habits, muntjac population sizes are difficult to assess and culling them can be a major challenge. Unlike the other deer species there is no close season for the shooting of muntjac. *The lack of a fixed breeding season means that female muntjac should be culled very selectively if the orphaning of dependant young is to be avoided*.

▶ The illegal driving of muntjac to shotguns for culling is regrettably still taking place. The shotgun, even in expert hands and used at an appropriate range, is seldom a humane instrument for the culling of any healthy deer.

▶ .22 centrefire rifles are now legal for the shooting of muntjac and Chinese water deer in England and Wales, provided specific muzzle energies and bullet weights are achieved. However, care should be taken in selecting an appropriate bullet for use on muntjac; fragile 'varminting' bullets may not expand slowly enough on impact to ensure adequate penetration and instead fragment on the exterior of the animal to wound rather than kill humanely.

▶ Muntjac venison, although excellent, has a limited market. These deer are relatively more time-consuming to process than others due to their tough skins, and some game dealers are reluctant to purchase carcases or offer a good price for this reason.

Chinese water deer (CWD)
Hydropotes inermis

↟ A mature Chinese water deer buck. Note the readily visible canine tusk.

Appearance

Similar in size to the muntjac, the CWD also stands about 50 centimetres high at the shoulder and an adult buck will weigh about 19 kilograms. However, it is less robust in build and is more typically deer-shaped with long legs and a delicate body; at a distance they may appear rather larger than they actually are and may occasionally be mistaken for roe deer.

The summer coat is a uniform light reddish-brown to sandy colour, with no distinctive facial or rump markings, becoming greyer and noticeably thicker in winter with a 'salt and pepper' appearance. The tail is short. The hind legs appear longer than the front ones, with the rump held higher than the shoulders. A combination of rounded, hairy ears and dark button eyes and nose give the face a 'teddy bear' look.

Unlike the other British deer species, the CWD buck never carries antlers. Instead he has well developed canine tusks, much larger than those of the muntjac and easily visible on the mature animal. Does also have canine tusks, but these are much smaller and only visible under closer examination.

Distribution

The CWD is our least widely distributed deer. It was introduced to Britain from China, like the muntjac, in the late 1890s and its establishment has been aided by escapes from zoos and parks along with some deliberate releases. It has not, however, been so successful in building numbers and spreading in the wild and is now largely confined to Bedfordshire, Cambridgeshire, Norfolk and Suffolk, though its range is expanding slowly. Other release sites may yet hold very small populations.

↑ Chinese water deer doe and fawn.

← British Deer Society Deer Distribution Survey 2016 results: Chinese water deer. Dark colours represent squares only filled in the 2016 survey, lighter colours squares reconfirmed in the 2016 survey and recorded in the 2007 and/or 2011 surveys, orange represents squares unconfirmed in 2016 but recorded in either the 2007 and/or 2011 surveys.

Habitat and feeding

In its native China, the CWD favours reed-beds and grasslands around the larger river estuaries; in the UK they have adapted well to parkland and wooded areas but still seem more at home in the more natural habitat offered by Fenland and the Norfolk Broads. Arable areas surrounded by thick hedges to retreat into are also acceptable to them.

Mainly a grazing deer, their predominant foodstuffs are young grasses, sedges and rushes, although they will browse on low tree branches and shrubs. They will also take crops such as carrots, potatoes and emerging cereals.

Behaviour

CWD are usually solitary, especially the bucks, but when food is short they may congregate to take advantage of good feeding areas. A typical feeding pattern involves grazing within a restricted area, then lying down to ruminate where they are rather than retire to somewhere more secure. They have a similar habit when disturbed, running only a relatively short distance before couching down in small patches of cover where they may still be visible.

Both sexes bark when alarmed. They are not as renowned for their jumping ability as other deer, but can get through very small gaps in fences and have something of a reputation as skilled escapologists in zoos and collections. They do not appear to interact much with the other deer species.

CWD are seldom implicated in damage to trees; they mark territories mainly by foot scrapes into which they urinate to deposit their scent. Scent marks are also placed directly onto foliage using small glands in front of their eyes.

They are not considered as resistant to British conditions as our other deer. The hairs on the winter coat are hollow and easily damaged. If this happens, the resulting loss of insulation can lead to hypothermia and death. Furthermore, CWD have a relatively short natural lifespan of around six years, which reduces an adult doe's breeding potential against that of other deer.

Breeding

CWD bucks become increasingly territorial as the rut approaches, and much fighting can occur. The canine tusks are used as weapons and injuries are commonplace, mostly to the neck and chest.

The rut takes place in December, and the young are born in May or June. Up to seven foetuses have been recorded in a doe, but two or three fawns are more normal. Mortality can be high if the birthing time and the weeks following it contain inclement weather, but the young grow fast, and female fawns are sexually mature in time for the rut. CWD does can therefore give birth on their first birthday, the earliest age of all the British deer.

Issues associated with CWD

❭ Of all our deer, the presence of CWD seems to be the most innocuous. Although they can cause localised damage to agricultural interests if their numbers are high, they generally have a low commercial or environmental impact.

❭ Whilst numbers remain low at the time of writing and their distribution is limited, one should not underestimate the CWD's capacity to reproduce quickly if conditions suit them. A succession of mild winters and clement birthing months could encourage numbers to rise rapidly, and significant further spread cannot be ruled out.

❭ Because it is difficult to distinguish between does and immature bucks easily due to a lack of antlers, the law permits only one open season for the shooting of both sexes. This allows only a restricted period for control where it is considered necessary.

❭ Despite a lack of antlers, the commercial stalking of CWD has a special value given its limited availability.

❭ Whilst CWD venison is excellent, their small carcase size does not allow for any large economic potential in this respect.

❭ It is worth noting that the British CWD population is considered to be taking on an international importance as the species declines in its countries of origin; Britain may contain one tenth, perhaps much more, of the world's numbers of wild-living CWD.[2] Whilst not currently an issue, this may yet have a bearing on our future management of the species.

2 Battersby, J. (Edited and compiled by), (2005), *UK Mammals: Species Status and Population Trends*, JNCC/Tracking Mammals Partnership 2005, ISBN 1 86107 568 5

Roe deer *Capreolus capreolus*

Appearance

Although significantly taller than the muntjac and CWD, about 75 centimetres at the shoulder, the roe is the third of what are usually classed as our small deer. A mature buck will weigh some 28 kilograms or more; the does are usually somewhat smaller. In poorer habitats body weights and overall sizes tend to decrease.

The roe is a graceful, leggy deer with large, mobile, black rimmed ears. It can be easily distinguished from other deer by the two distinct white spots at the end of the muzzle above the upper lip and a white chin. Unlike the other deer species, there is no tail visible on the live animal although it will be found as a short stump when handling a carcase. The summer coat is a rich foxy red, becoming duller and darker in winter. Some animals may have one or two white 'gorget' patches on their throats.

The caudal patch is white and made up of erectile hair which can be flared if the animal is alarmed; this is especially conspicuous when the animal is in its winter coat. The doe also has an anal tush, a tuft of hair which looks like an upside-down shaving brush, at the base of the caudal patch – this is also a key identification point, especially useful during winter culling operations, as neither roe bucks nor other deer have such a feature.

Bucks grow simple antlers from short pedicles on the skull. Six points (three each side) are typical on a mature animal, although abnormalities are not uncommon and can result from a number of causes. Antlers are shed around November or December and regrowth starts immediately. They are hardened and clean of their protective velvet by around March or April. At such times the buck can cause damage to saplings as he frays them vigorously to remove the velvet.

Distribution

The roe is probably the most widespread of our deer and is found throughout mainland UK, though it is less common in some parts of central England and Wales.

A roe buck in spring, showing the change from winter coat (rump and flanks) to summer coat (shoulder and neck).

Roe doe in summer coat.

It is not present in Ireland (where an introduced population once existed but has long since died out), nor on a number of Scottish islands or the Isle of Man.

Habitat and feeding

Though they are predominantly a woodland deer, roe are remarkably adaptable and can take up residence in a wide variety of places, provided these offer them shelter and good feeding. 'Field roe' are increasingly encountered, living the majority of their lives in open areas, as long as there are surrounding hedges for lying up when not out feeding. They will also use moorland widely, although such animals tend to be smaller than their woodland counterparts, reflecting the poorer habitat. Such animals can be very difficult to approach as they are quick to spot any movement and take full advantage of the security offered by wide fields of view.

As ruminants, like all deer, roe chew the cud and are concentrate selectors who rely on a limited quantity of high quality food. Hardwood coppice, brambles, buds and newly flushed leaves are all taken, as well as any seasonal new growth. The overall quality of the forage available has a direct impact on the extent of an animal's range or territory, as indeed it does on the local density of populations.

As late autumn and winter approach, roe will move and eat less to conserve their energy against the approaching harder months. Their metabolic rate drops, and their food intake can reduce by as much as 40%. This usually occurs around the time that the roe doe season opens, leading many observers to believe that the deer have disappeared from their ground. By Christmas the deer tend to start to move more as they are forced into foraging by decreasing internal reserves.

Behaviour

Another solitary species, the roe is more usually seen alone or in small family groups throughout most of the year. The bucks are very territorial during spring and summer. They fray small trees or saplings with their antlers, often leaving a small scrape made with their forefeet at the base of a fraying stock. At this time they are very aggressive towards each other and much barking and chasing occurs. Fighting only tends to take place where two bucks are evenly matched and a defeated buck, if prevented from escaping its rival, can be badly gored by the victor's antlers or even killed.

Both sexes bark, a deeper sound than that of the muntjac and not as persistent. Barks can serve as an alarm to alert other deer to the presence of danger, as a challenge or a warning. Roe also make a variety of other noises, predominantly squeaks, which are often imitated to attract rutting bucks or the does leading them.

During winter, with the falling off of testosterone levels, roe will become more tolerant towards each other and are often seen taking advantage of areas with better feeding in large groups. These are not herds, as there is no real social structure within them, and if disturbed the group will usually disperse in a number of directions.

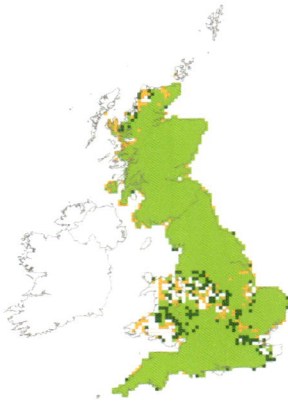

↑ *British Deer Society Deer Distribution Survey 2016 results: Roe deer. Dark colours represent squares only filled in the 2016 survey, lighter colours squares reconfirmed in the 2016 survey and recorded in the 2007 and/or 2011 surveys, orange represents squares unconfirmed in 2016 but recorded in either the 2007 and/or 2011 surveys.*

Breeding

The roe rut normally takes place from around the end of July and into the first half of August. The doe attracts a buck to her by calling and leaving scent marks, and mating takes place after prolonged pursuit by the buck. 'Roe rings', flattened circles or figures-of-eight in the ground foliage of the woodland floor, can occasionally be found after such pursuits.

The roe is unique among all deer in that it exhibits a process known as 'delayed implantation', i.e. the fertilised egg does not implant into the wall of the uterus and start to develop properly until December. This allows the deer to rut and still recover condition well before winter, yet still give birth at a time of good weather and abundant forage.

Kids, usually twins, are born in May or June. Triplets are not uncommon in good breeding years, though a small doe or one in poor condition may give birth to just a single kid. Offspring remain with their mothers for most of their first year until she drives them off shortly before giving birth again.

Issues associated with roe

◗ Fraying damage is a perennial roe problem, starting early in the year when the bucks clean their antlers of velvet, then later as they mark out and maintain territories. Fraying stocks are often in conspicuous places, sometimes making the problem seem worse than it actually is. Fraying damages tree bark, either killing the plant or stunting its growth.

◗ The browsing of new shoots can cause a young tree to grow erratically, reducing its timber value when fully grown.

← *A roe kid pictured in early August, aged around three months. His juvenile spots have almost faded and the small bony protrusions on his head (pedicles) from which antlers will eventually grow are starting to develop.*

▶ Young bucks, unable to hold territories, may congregate in 'coffee bar' areas where they cause significant fraying damage through competing with each other. For this reason, it is often wiser to accept the lesser damage caused by a dominant buck than to cull him and permit larger numbers of lesser bucks in.

▶ Roe are especially intolerant of over-population; too many animals can cause stress and build-ups of disease. When food is short, significant winter die-offs have been known in places where roe numbers outmatch the carrying capacity of the habitat.

▶ Winter 'herds' of roe congregating at good feeding areas can cause significant damage to emerging or winter crops.

▶ The doe cull is often hampered by the winter inappetence which limits their movements, and thus their visibility, for the first two months of the open season. Conflict with game shooting interests may also restrict culling activities until pheasant shooting finishes at the end of January, leaving only two months to complete culls. Some stalkers may become reluctant to cull mature does in February and March because they will be carrying increasingly large foetuses.

▶ Captive roe bucks have a reputation for unpredictable behaviour, especially if hand-reared, and can be highly aggressive towards humans. For this reason, roe are rarely seen in zoos and collections.

▶ Roe bucks have a high sporting value, and there is particular demand for roe stalking from continental sportsmen who recognise the high quality of trophies available in Britain. Roe venison has a distinctive quality and enjoys a ready market.

Fallow deer *Dama dama*

Appearance

The fallow is the first of our three large deer species. A mature buck will stand almost a metre high at the shoulder and weigh as much as 100 kilogrammes; does are about 10 centimetres shorter and weigh around 55 kilogrammes.

Fallow occur in four main colour variations:

Common – the coat is a rich chestnut colour with cream spots in summer, changing to a two-tone grey/brown in winter, with the spots having all but disappeared. The rump patch is white and surrounded by a black marking resembling an inverted horseshoe. The tail has a thick black stripe down it.

Menil – the coat is a light beige in colour with white spots; the change to the winter coat dulls the colours somewhat but the spots are still visible. Like the common variety, the rump patch is white but the markings surrounding it and on the tail are brown rather than black.

Black - the glossy black summer coat shows few signs of spots. In winter, it becomes duller and the lower half of the body is a dark mushroom colour. The tail and rump patch are black.

White - a creamy white all year round, with pale noses and hooves. This colour phase is not albino as the eyes are dark.

A mature fallow buck of the menil variety. A young buck with his first set of antlers is in the background.

A mixed group of the different fallow colour varieties, including a well-grown calf.

White fallow lack the pink eyes of a true albino.

The tail is long, for a deer, and very mobile. Fallow have a relatively long face and broad muzzle; bucks have a prominent 'Adam's apple' and the guard hairs on the penis sheath show as a tassel. Of all the British deer, adult fallow have the most sophisticated antlers which tend to be wide, flattened and palmated although bucks do not grow a fully mature set until they are about six or seven years old. Antlers are usually shed around April or May, and are regrown and clean of velvet by September.

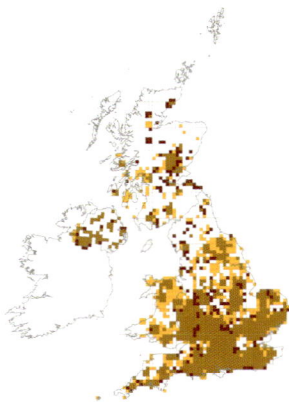

↑ British Deer Society Deer Distribution Survey 2016 results: Fallow deer. Dark colours represent squares only filled in the 2016 survey, lighter colours squares reconfirmed in the 2016 survey and recorded in the 2007 and/or 2011 surveys, orange represents squares unconfirmed in 2016 but recorded in either the 2007 and/or 2011 surveys.

Distribution

Introduced into Britain by the Romans and then more extensively by the Normans, wild fallow now occur throughout England and Wales, and are slightly less widespread in Scotland and Northern Ireland. They are, however, a deer that largely confines itself to traditional ranges and habitats and can be abundant in one area whilst never seen only a few miles away.

Habitat and feeding

Whilst deciduous woodland is the most natural habitat to fallow, they will adapt readily to any area which offers them cover and security. The dappled coat of the common fallow is a particularly effective camouflage in the broken light of woodlands. They will make heavy use of arable land adjacent to woods, particularly where it offers thick hedgerows which prevent the deer from feeling too exposed to view.

Although they have the wide muzzles of a grazing species, fallow will also browse. They eat a wide variety of plants according to season. Following the first flush of new leaves and buds, the main staple throughout spring to autumn is sweet grasses, sedges and rushes. Herbs are an important summer food and, in areas of high deer densities, a browse line marks feeding on the leaves and shoots of broad-leaved trees and shrubs. Grasses and herbs of less than four inches in height are preferred; once they have grown higher than that, the deer tend to move elsewhere to feed. As the winter months approach, acorns, beech mast, crab apples and chestnuts form a significant part of the diet. Where the nut crop is especially poor, holly, bramble and ivy will be eaten readily (the latter two are especially important as a winter food source).

Arable crops are taken readily, as are winter vegetables. Game crops such as mustard, canary grass, kale and maize will also attract fallow where they are available.

Behaviour

Fallow are a herding deer species and tend to spend most of the year in single-sex groups. Does in particular can form very large herds, usually led by an older animal, and their male offspring may remain with them for up to two years. During the summer, smaller groups are often encountered. Herds tend to have traditional ranges and to stay within them unless forced elsewhere by pressures such as excessive disturbance or food shortages. Where they are hunted intensively, fallow will become almost entirely nocturnal in their habits and seldom be seen during the hours of daylight.

Pronking is a gait especially associated with fallow, in which all four feet take off and land together. Although this is not as fast a pace as galloping, the pronking deer is highly visible to others as an alarm signal.

Fallow are characteristically silent throughout most of the year, although during the rut a buck makes a continuous belching snort as he patrols his rutting area; this is often referred to as 'groaning'. Otherwise both sexes may utter a single bark as an alarm signal; fawns and does communicate with each other by bleating.

Breeding

The fallow rut reaches a peak in late October, but may start as much as a month earlier. The bucks, by now in peak condition and with visibly swollen necks, move to their traditional rutting areas where they establish lekking stands by scraping the ground with their hooves, fraying and thrashing foliage with their antlers, and scent-marking. They can also exude a strong smell, often described as being like rancid butter. The does gather around the rutting stands, and intense fighting can take place between competing bucks. A loss of weight and general decline in overall condition becomes apparent among the dominant bucks as the rut progresses.

Single fawns are born around June, although late fawning is quite common. Often a doe will allow a fawn which is not her own to suckle, giving a false impression that she has given birth to twins. The birthing of twins is, in fact, very rare in all of our large deer species.

Issues associated with fallow

▶ Large herds of fallow can cause intense levels of agricultural damage. Emerging crops are taken, and animals will roll in and flatten established grain crops. Winter roots are also attractive to them; lacking upper incisors like all deer, single plants are often uprooted to leave a single bite before the deer moves on to the next one.

▶ Many stalkers consider fallow the most difficult deer to control effectively. Large herds, with many sets of suspicious eyes, are more difficult to approach within shooting distance. Furthermore, fallow learn quickly and are very reactive to excessive stalking pressure, varying their habits accordingly and sometimes then only moving by night.

▶ Seasonal movements between feeding and rutting areas may mean that a herd can range across various properties. A lack of co-ordination between landowners can lead to ineffectual management of numbers, with some parties gaining unequal opportunities for the more attractive and lucrative buck stalking according to where the deer may be congregated at a specific time of year. In many cases a co-operative approach is essential.

▶ Some commentators consider fallow to be the most mismanaged deer species in Britain. There is no doubt that mature bucks are over-shot in many instances, to the detriment of the more important reduction of doe numbers. This is not helped by the fact that bucks remain in season throughout, and beyond, the doe season, and offer more attractive trophies as well as more valuable carcase weights for venison sales, thus distracting some stalking effort.

▶ Road traffic accidents are a major issue in some areas. As a herding deer, motorists are often unaware that once a fallow has crossed in front of them others may be following. Although smaller species of deer are also found in areas where heavily used roads are frequently bordered by woodland, the greater height of a fallow makes it more liable to tip over a car bonnet when hit and subsequently through the windscreen. It is also possible for the animal to be projected by the impact into the path of another vehicle. A higher body weight makes such an impact even more likely to be catastrophic to the driver.

Sika *Cervus nippon*

Appearance

Very similar in size to fallow, a sika stag stands some 95 centimetres high at the shoulder, while a hind is about 10 centimetres smaller. A mature stag will weigh 80 kilograms or more.

The summer coat is a chestnut brown, covered in creamy white spots, shading into a grey neck and with paler underparts. Sika are sometimes mistaken for fallow at this time. In winter the coat becomes a more uniform charcoal grey; in the case of mature stags, it may be so dark that it appears to be black.

The rump is pure white with a dark border throughout the year, and comprises of erectile hair which can be flared out as a signal to others that the animal is alarmed. The tail of the sika is white and usually has a thin black stripe down it; it is not as long or as mobile as that of the fallow. The metatarsal glands on the hind legs are very prominent, and appear as a pale or white oblong.

↑ *Sika stag in winter coat, with typical eight point antlers and traces of the summer spotted coat remaining.*

➡ *A group of sika hinds in summer coat, chestnut brown with visible spots.*

Both sexes have a dark U- or V-shaped marking on the forehead that gives them a 'frowny' expression. The antlers of a mature stag normally have no more than eight points (four on each side) – any more than this is abnormal.

Distribution

The sika is another non-native species which was introduced into Britain from the Far East at around the same time as the muntjac and CWD. Again, most of our wild populations derive from escapes or deliberate releases. In England, the main herds are in Dorset, the New Forest and Cumbria and Lancashire, with a few smaller populations elsewhere. Localised herds also exist in Northern Ireland.

In Scotland they continue to expand their range rapidly, no doubt aided by the large coniferous forests and peaty soils which are a preferred habitat feature.

Habitat and feeding

Sika are an adaptable deer. Initially seemingly constrained by areas characterised by acidic soils, they are now appearing regularly outside their preferred heathlands and conifer forests as expanding populations have encouraged them to move elsewhere. They are secretive, however, and unwilling to settle in large open expanses without suitable retreats.

Their feeding habits, like those of the fallow, are cosmopolitan and sika are predominantly grazing animals which will also browse. They readily adapt to whatever is available according to season. They are quite capable of thriving on rough, coarse forage material.

Behaviour

Another herding species, mature sika tend to separate into single sex groups throughout most of the year. Juvenile stags usually join the main stag herds at about a year old, although sometimes younger than this. During the summer months, single hinds are often found with just their calves, and possibly sometimes a yearling animal which is probably their calf from the previous year. Once the rut is over, these small groups generally join up into larger herds.

Recent studies indicate that sika seem to 'heft' to specific areas from which they are reluctant to stray far, selecting their home ranges according to landscape structure and actual habitat availability.[3] Young stags may, however, travel longer distances and are usually the first sika to turn up in areas where they have not been noted before. Both sexes are secretive and tend to stay close to cover until after dark.

Sika stags are very vocal and have a wide range of calls. During the rut they make a rising and falling whistle, often repeated several times, which can be an eerie sound you would not normally associate with deer. They also grunt and snort at rivals. Hinds tend to be rather quieter, but if alarmed will emit a loud and piercing squeak which may be repeated, especially if the source of potential danger remains suspected.

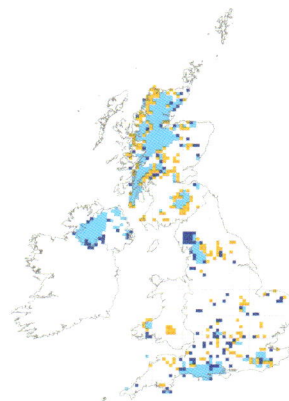

↑ *British Deer Society Deer Distribution Survey 2016 results: Sika. Dark colours represent squares only filled in the 2016 survey, lighter colours squares reconfirmed in the 2016 survey and recorded in the 2007 and/or 2011 surveys, orange represents squares unconfirmed in 2016 but recorded in either the 2007 and/or 2011 surveys.*

3 Uzal, A., Walls, S., Stillman, R.A. and Diaz, A. (2011), *Sika deer distribution and habitat selection: The influence of the availability and distribution of food, cover, and threats*, European Journal of Wildlife Research Volume 59 Issue 4

➡ *Sika (pictured) and red stags both develop a pronounced mane during the rut.*

Breeding

The stag groups will break up at the end of the summer in preparation for the rut, and stags demonstrate a variety of rutting behaviours. Territories are marked by scrapes, wallows, thrashing trees with their antlers, and scoring tree trunks with their antlers (bole scoring). Usually a stag will round up hinds and attempt to hold them in his territory, but where hinds are plentiful stags may set up lekking stands, with several in a limited area, and try to attract the hinds to them.

Even where densities are high, a large proportion of mature and yearling hinds tend to mate successfully. In good years, where a calf grows quickly enough to reach a breeding body weight in time for the rut, it may breed as well. The single calf is born in May or June, although later births are not uncommon. Calves grow rapidly and are almost the same height as their mothers by their first winter.

Issues associated with sika

❯ Sika and red deer, being closely related, are capable of hybridising and producing fertile offspring. This does not appear to be an issue where populations of the two species are properly balanced, but by the 1970s hybridisation was being noted among wild populations.[4] The problem has since grown to the point that, at the time of writing, there are already distinct fears for the genetic integrity of the red deer in parts of mainland Scotland; in one area studied recently, more than 40% of deer were found to be a mixture of red and sika.[5]

❯ Large sika herds can cause significant levels of agricultural damage. Bole scoring (where the trunks of trees are gouged by the antlers of stags) is a major form of forestry damage specifically associated with sika.

4 Horwood, M.T. and Masters, E.H (1970), *Sika Deer*, 2nd edition, Warminster, The British Deer Society

5 Pemberton, J.M. and Senn, H.V. (2009), *Variable extent of hybridization between invasive sika* (Cervus nippon*) and native red deer* (C. elaphus) *in a small geographical area*. Molecular Ecology, 18: 862-876

▶ Although sika are more usually attracted to areas of acid soils, such as coniferous forestry and heathland, they are beginning to show signs of adapting to other areas where their numbers have increased to high levels and they are forced to seek new habitats. Dorset sika, for example, are already moving into the chalklands which surround their more usual haunts.

▶ Sika hinds tend to produce calves annually, unlike red hinds which may only produce young in alternate years in areas where habitat and conditions are not conducive to more sustained breeding (such as some parts of Scotland).

▶ Of all the British deer, sika stags have a reputation for aggression when wounded and have occasionally been reported as attacking, rather than fleeing from, stalkers and dogs approaching them. For this reason it is strongly advised that wounded sika should be finished off with a rifle shot from a safe distance.

▶ There is no doubt that sika are particularly robust and should be shot with an adequate calibre of rifle. If the bullet does not deliver sufficient kinetic energy for an effective knock-down on the spot, they can be liable to 'carry the shot' even though technically dead and make off to collapse in thick cover or equally difficult locations for extraction.

Red deer *Cervus elaphus*

Appearance

Probably the most iconic of our British deer, the red is certainly the largest as well. Size varies considerably according to habitat however; a mature woodland stag in the West Country or East Anglia can stand almost 140 centimetres at the shoulder and weigh well over 200 kilograms. Its smaller hill counterpart in the Highlands of Scotland may be only slightly taller than 100 centimetres and weigh only 120 kilograms. The weights of hinds, and even the birth weights of calves, can show a proportionally similar set of variations.

The summer coat, as their name suggests, is a rich red colour with a lighter underside becoming greyer in winter. There may be some slight regional variations. The rump patch extends above the base of the short tail, and ranges between lemony yellow to off-white. Like sika, the stags develop thickened necks and a shaggy mane during the rut.

Antlers vary considerably according to the age of the stag and the habitat available to it. Each year, as the antler is cast and regrown, it will normally increase in length and thickness until the animal reaches its prime at anywhere between eight and twelve years old. Woodland stags tend to develop the more impressive multi-point heads; an average hill red deer in poorer habitat may never exceed a total of eight or ten points. An important distinguishing feature between sika and red antlers is that the bottom antler point, or 'brow tine', usually leaves the main beam at an angle of about forty-five degrees on the sika and ninety degrees on the red.

➡ *A red stag proclaims his mastery during the rut.*

⬇ *A red deer hind and her calf, pictured in late summer.*

Red stags without antlers, known as hummels, are not uncommon. This condition is caused by a poor start in life which prevents the calf from developing the bony pedicles on the skull from which the antlers eventually grow. Subsequent resources are thereafter diverted into developing body weight rather than antler growth, and the mature hummel can be a large animal which frequently breeds successfully. The condition is not, however, hereditary.

Distribution

Red deer are native to Britain and have been resident here since the Ice Age. They are widely distributed across the British Isles, although the majority are found in the Highlands with large woodland herds in southern and western areas of Scotland. In England, the main populations are in Cumbria, Yorkshire, East Anglia, the New Forest and the West Country. Elsewhere there are smaller populations, usually based on initial escapes from park herds.

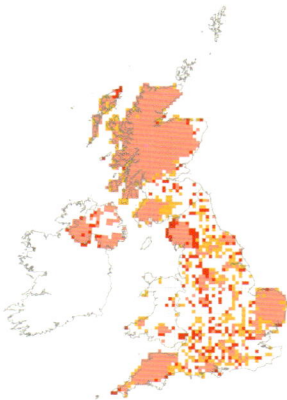

⬆ *British Deer Society Deer Distribution Survey 2016 results: Red deer. Dark colours represent squares only filled in the 2016 survey, lighter colours squares reconfirmed in the 2016 survey and recorded in the 2007 and/or 2011 surveys, orange represents squares unconfirmed in 2016 but recorded in either the 2007 and/or 2011 surveys.*

Habitat and feeding

By nature a woodland or forest creature, the red deer has been forced to adapt in some places, with surprising success, to less hospitable environments such as the Highlands of Scotland. This adaptation has been marked by smaller body sizes and behavioural differences.

The red deer is a bulk feeder which will both graze and browse, taking whatever suitable foodstuffs are available. On the open hill this may amount to little more than heather and grasses; in richer country there are better choices available to them. In leaner times they will strip bark from trees in quantity, and agricultural land is frequently raided for winter crops. A disproportionate level of damage can occur among root crops, such as beet or turnips. As with all deer, red deer have no upper incisors so, instead of consuming the whole root, the first bite often pulls it from the ground where it is left as the animal simply moves to the next one to do the same thing again.

Behaviour

In common with the other large species, red deer are herding animals which tend to live in single sex groups for most of the year. The size of these herds can vary greatly; woodland reds seldom form parcels of much more than a dozen animals, whilst on the open hill it is quite normal to see more than a hundred hinds, often many more, in a herd. The range covered by the herds also varies accordingly.

Definitive 'pecking orders' within herds are established, especially amongst the hinds, which will box with their forefeet to establish dominance. Stags, wary of protecting growing antlers, will also do this until the latter are fully grown and ready for proper use.

Both sexes like to wallow. Outside the rut, the clinging mud provides some protection against biting insects, whilst during the rut stags wallow to enhance their physical appearance. Red deer swim readily and well, and even large expanses of water do not seem to deter them.

Breeding

As the rut approaches, usually around mid-September, the stag groups separate and individual animals seek to round up and hold hinds, protecting them against rivals. The stags roar in challenge to each other, and very often the sound of a stronger stag is sufficient to deter a rival. For this reason hummels, even though they lack antlers, can still be successful in holding hinds and breeding. Where animals are evenly matched, however, fighting can occur, with two stags locking antlers in a ferocious pushing match. The victor is usually the heavier animal, not necessarily the one with the more impressive antlers.

During the rut stags are too busy constantly herding their hinds and protecting them against rivals to feed. As a result they can lose as much as 40% of their body weight and are eventually unable to see off rivals, who will then take over their hinds and mate with those that have not already come into oestrus.

⬇ *A dominant red stag with his harem of hinds.*

49

Calving takes place in June, although earlier or later birthings are not uncommon. The single calf is born away from the main herd but is soon mobile, and is taken by its mother to join it once strong enough. A hind that fails to regain condition quickly after calving may spend the following year barren and is known as 'yeld'. This is a more common occurrence in poorer habitats.

Issues associated with red deer

▶ Attempts to improve the quality of smaller 'hill' red deer in Scotland by introducing larger, woodland animals, or even American wapiti (a larger, closely related species) have invariably failed. Even if there is an initial gain in body weights, they will diminish through subsequent breeding to the size that the habitat can naturally support.

▶ As noted in the previous section, red deer are closely related to sika and hybridise readily where conditions allow. Mismanagement, particularly through the overshooting of mature red stags, creates population imbalances and stresses which enable such conditions; unaccompanied red hinds will be encountered by sika stags which will successfully breed with them. Given the marked difference in size, it is unlikely that red stags will mate easily with sika hinds.

▶ It should be noted that hybridisation is not inevitable when proper population dynamics exist. Red and sika co-exist in many deer parks, with their separate ruts taking place in close proximity to each other without any interaction between the species.[6]

▶ Large herds can be associated with very high levels of agricultural damage throughout the year. In the north of England and Scotland this can be especially acute during the winter months, as animals are forced to descend onto lower ground by inclement weather and a lack of natural forage.

▶ Red deer will habitually strip bark from trees in forestry and high levels of damage can result.

▶ Red deer are large animals and carcase handling has special implications. The stalker, especially if working alone, will require specialised mechanical handling equipment to cope with this and storage facilities will need particularly careful design regarding lifting gear, volume and load-bearing.

▶ As trophies of the chase, red deer antlers (even those of lesser quality) can have a high trophy value to sportsmen.

▶ There is a high demand for red deer venison. However, care should be taken to ensure that the carcases of red stags shot towards the end of the rut are not passed into the human food chain. Such 'spent' animals are in poor condition and the meat quality is accordingly of a very low grade. Many consumers will judge venison by such poor experiences and may be unlikely to return to it; this does the venison market, and the deer manager looking for a return from it, no favours. ■

6 Prior, R. (1987), *Deer Watch*, 1993 edition, Shrewsbury, Swan Hill Press

The British deer species can be easily distinguished by their different rump markings and comparative tail lengths:

← **Sika** – medium length tail, white with thin black stripe down centre. Black edging to pale rump, extending down to hocks. White erectile rump of hair can be puffed out if alarmed.

↑ **Fallow** – long tail with thick stripe down centre and inverted horseshoe marking around edge of pale rump. Markings are black on common variety, brown on menil, not visible on black or white.

➜ **Red** – short tail, pale rump patch (shades can vary) which extends above base of tail.

➜ **Muntjac** – medium length, thick tail with white underside showing on edge.

↑ **Roe** – no visible tail, white rump patch of erectile hair (more obvious on winter coat). Only the roe doe has an anal tush, a downward-projecting lump of hair at the base of the rump patch (pictured).

← **Chinese water deer** – very short tail, no rump markings.

51

3 Management: Passive or Active?

The reaction of many people, on discovering that they have a problem with deer, is to pick up a gun or to contact someone who has one. Yet there are alternative approaches which may be even more effective, albeit sometimes only as a stop-gap solution. Many deer issues are seasonal or short-term anyway, such as the need to prevent damage to emerging crops, to protect young trees until they have grown beyond the point that they are attractive to deer, or simply to keep deer away from sensitive areas for a specific period.

Indeed, there are places where shooting is inappropriate and should not even be considered. When deer are causing a problem in a built-up area, or indeed close to any form of human habitation, the use of firearms may well be precluded for reasons of safety, noise and a desire not to alarm or outrage the neighbours.

Passive protection methods – that is, excluding, deterring, repelling or protecting against deer without taking more direct action against them – are often all that may be needed. There is one caveat though. In the face of growing deer numbers, you may just be shifting a problem away from one place and on to another. Whilst wild deer are the property of no one, this is likely to be a poor explanation to a neighbour who is suffering damage because deer have relocated from your land onto theirs.

Sometimes, then, a more active form of management is called for. In this chapter we will examine the various options available across the whole spectrum of passive and active deer management.

Identifying the culprit

Where damage occurs, all too often the finger is pointed at deer without first considering whether it may be attributable to another cause. Before rushing in and expending time and energy – and quite often a great deal of money – it is worth examining the evidence to check whether deer really are to blame. One call out to a badly damaged field of standing maize identified badgers as the real culprits; another, from a forester enraged that deer had seriously damaged an area of new plantation, took little examination before hair and copious cowpats made it obvious that cattle had broken through the fence.

Check for physical signs of what might have caused the damage – tracks and droppings are a good start. Deer will leave their tracks in soft ground, being heavier

than rabbits or hares whose tracks are only really likely to show up in snow. It is easy to distinguish between the droppings of the different species; those of deer tend to be dark brown or black, smooth and glossy in appearance, oval in shape and usually have a small indent in one end. Those of rabbits and hares are rounded and very fibrous.

If damage to saplings or crops is being investigated, look carefully at the bite marks. Deer have no upper incisor teeth with which to shear their food plants. Instead they grip them between a pad of skin on the upper palate of the mouth and the lower incisors, half cutting and half tearing the shoot. If a rabbit or hare has done the damage a cleaner cut is apparent.

The fraying of trees and bushes tends to occur for two main reasons. It is caused by antlered species which are in the process of cleaning velvet (the skin which covers, nourishes and protects the developing antlers) when growth is complete. It is also an important means of communication between male deer as they mark their territories. Significant levels of damage can be caused, especially by the larger species which may also thrash vegetation indiscriminately as a display activity.

1 *Red deer droppings (with 50p piece for scale) shiny, smooth, ovoid and usually indented.*

2 *Rabbit droppings - rounded and fibrous.*

3 *The diminutive footprint or 'slot' of a muntjac.*

4 *The bite signature of a deer. Having no upper incisor teeth it will show on the plant as half cut, half tear.*

5 *A hare or a rabbit, having both upper and lower incisors, will shear more cleanly through a plant stem.*

6 *Bark stripping by rabbits or hares is usually on fallen branches or at the very base of trees, with teeth marks often clearly visible.*

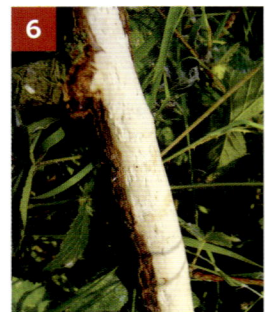

Maximum browsing heights		
Species	Normal	Standing on hind legs
Chinese water deer	0.6m	1.0m
Muntjac	0.6m	0.9m
Roe	1.15m	1.5m
Fallow	1.4m	1.7m
Sika	1.4m	1.8m
Red	1.5m	1.9m

Fraying heights and times		
Species	Height	Time of year
Chinese water deer	0.5m	Bark scraping with tusks can occur mainly in November and December around the time of the rut but may also occur at other times of the year. Effects are not as significant as for other deer species.
Muntjac	0.2 to 0.5m	May occur at any time of year due to lack of seasonal rutting cycle; canine teeth used in addition to antlers.
Roe	0.8m	February to May for velvet removal, April to mid-August as territorial marking/display.
Fallow	1.6m	July to August for velvet removal, mid-September to October as territorial marking/display.
Sika	1.6m	July to August for velvet removal. Fraying is not so important to sika as it is to other species for territorial marking, but the fraying of trees and bole scoring around rutting areas may occur September to mid-October.
Red	1.8m	Mid-July to mid-September for velvet removal, then to late October as territorial marking/display.

Just because a deer was seen in the area under attack, it doesn't mean that it was the main culprit. At a distance a deer can be very obvious - a rabbit, or even a small horde of them, is much smaller and may not stand out so much!

Passive Approaches: Exclusion

Fences

Any visitor to a historic house or stately home will probably have come across a walled garden, constructed with the main aim of keeping four-legged intruders away from precious plants and vegetables. Building a wall today would probably be prohibitively expensive for most people but you still have to face it; the only really effective way to protect a garden or other sensitive area from deer is to exclude them, and I'm afraid that a proper deer-proof barrier is no half-hearted affair. For a start, it has to be high enough. Deer are excellent jumpers: as a general rule of thumb, if a deer can stand on its hind legs and reach the top of a fence with its chin, it can jump it. Even small deer jump well; muntjac have been known to leap onto a sheer, five-foot-high bank from a standing start with apparent ease.

Standard stock fencing is too low to exclude deer and actually presents special dangers to the animals which, in jumping them, can twist a leg (usually a hind one) between the top strands and become trapped. For this reason, only a single strand topping for stock fences is recommended.

In the past a walled garden was often built to protect valuable plants from the depredations of deer and other unwanted visitors.

Photograph: A J Parker

← *Stock fences can present a hazard to deer jumping them. This unfortunate young sika stag, which had to be euthanised, was caught between the top two strands.*

Your fence needs to be robust enough to stop a deer tangling legs, antlers or body in it, with a sufficient gauge of mesh to prevent the animal from squeezing through; if the head fits, you'd be surprised at how easily the body can follow. Flimsy materials, such as chicken wire or light plastic or nylon mesh, should be avoided. Even the smallest breaches, often created by badgers, are quickly exploited and widened. The Forestry Commission is an excellent source of information on fence design.[1]

I recall visiting some local allotments which, despite a fine high fence, were still being raided by roe. The owners were mystified until I was able to point out

← *Inadequate fencing. A gap originally created by badgers has been enlarged by fallow.*

numerous creeps the deer were using to get underneath it. Given the option, a deer will nearly always choose to go under, rather than over, an obstacle. As a result the base must be securely pegged or, better yet, buried or folded at a right angle on the outside and turfed over. Again, the smallest gap will give the deer an opportunity to get through. On one occasion, I observed a roe doe and her well-grown kids travelling beside a high chain link fence when they disappeared from view behind some bramble. When they reappeared they were on the far side of the fence. After the trio had moved off I examined the short section where they had been out of

1 Trout, R. and Pepper, H. (2006), *Forest Fencing*, Forestry Commission Technical Guide

sight. The only place where they could have got through was a small gap a few inches high where the base of the fence did not meet the ground. A few shed hairs confirmed that it had been used.

➜ *Gaps at the base of a fence will be exploited; this is a regular roe creepway.*

Recommended minimum fence and tree guard dimensions. These should offer at least 95% effective protection against deer:		
Species	Fence height	Fence mesh size
Muntjac	1.5m	75 x 75mm (while 100 x 100mm mesh is usually proof against muntjac, bucks can occasionally get their heads through and snag their antlers.
Roe	1.5m	200 x 150mm
Fallow & Sika	1.8m	220 x 200mm
Red	1.9m	

As a proper deer fence is expensive (you may be looking at around £10 per metre at current 2019 prices) and unsightly, this option may not be for you, so what are the alternatives? Two rather more simply constructed low fences, or a fence combined with a hedge with a narrow strip of grass between them, have been known to deter deer from jumping into such a restricted area. As a general rule, deer do not like to feel confined or restricted in their movements.

Electric fencing, either using wire strands or netting, can be helpful if circumstances allow, but do take care if you have pets or children. The wires need to be highly visible and the deer will quickly learn to go over or under them, so a low, stand-off wire will help to deter them. Electric fences, of course, must be regularly checked and maintained if they are to remain effective. Roe deer are said to be more tolerant of electric shocks than the larger deer and the electric fence may accordingly be less effective in excluding them.[2]

Trapped deer

▶ Occasionally deer will become 'trapped' in fenced areas. If left to their own devices and undisturbed they will usually find their way out the same way they came in. Leave any gates or other exits open and available to them.

▶ Do not be tempted to try and herd a deer out of an enclosed area. Inevitably, it will panic and run the risk of damaging itself as it attempts to escape.

▶ Smaller deer, such as roe and particularly muntjac, can become trapped by the body in wider mesh fences. Approaching them from the front will often encourage them to back out the way they got in without injuring themselves.

▶ If a deer is firmly caught in a fence, the combination of its own attempts to free itself and the trauma caused by the close proximity of human rescuers may result in unacceptable stresses and serious injury. In such cases, euthanasia may be a more humane option than release.

Deer leaps that offer access from high to low ground, but not vice versa (in effect a sort of reverse ha-ha), can be constructed within fence lines to allow deer that have become inadvertently fenced in to exit enclosed areas whilst not being able to re-enter them. These are especially useful when accidental incursions – such as when snowdrifts have built up enough to allow deer over the fences in winter – are likely to occur. They need to be sited according to local conditions and the topography of the ground, and it may be necessary initially to put food down to encourage the deer to use them. Otherwise deer tend to be naturally inclined to follow fence lines and will have to discover them for themselves.

Essential gaps in fences, such as those needed for vehicle access, must be provided with reliable and effective gates, the closing of which must be rigorously enforced if the fence is not to be rendered useless. Cattle grids may provide an alternative to gates but need to be wide enough to prevent deer jumping across them: 4 metres should discourage most species, but some manufacturers recommend that as much as 4.6 metres may be necessary where red deer are present. They should have rounded rather than flat upper surfaces as deer have been known to learn to negotiate the latter. The space under any grid must be well drained as freezing weather may create ice which an animal will then simply walk across, and should also be provided with an exit ramp for any small animals which fall through and might otherwise become trapped (hedgehogs are especially prone to this).

2 Prior, R. (1995), *The Roe Deer*, Shrewsbury, Swan Hill Press

Individual plant protection

If you have particularly valuable plants and shrubs you can consider protecting them individually, if only until they are established. Tree tubes, though unattractive, can be effective if securely staked and a suitable height, and more decorative guards are available – at a cost. Otherwise, standard tubes come in a variety of finishes, ranging from rigid plastic to spirals or open-sided heavy mesh.

Tree tubes can be a very popular option for plant protection, both in the garden and over larger areas of young plantation. The tubes provide plants with an improved microclimate to give them the best possible start, whilst encouraging growth and increasing survival rates. They offer good protection against animal damage and will also help to ensure that herbicide does not come into accidental contact with plants during spraying.

To be fully effective, the tubes need to be of the correct height to protect against the species present. If not, all you are doing is providing the deer with a 'breakfast bar' which allows growing shoots to be browsed as soon as they emerge from the top of the tube. Some recommended heights are below:

Species	Tree guard height
Muntjac	1.2m
Roe	1.2m
Fallow & Sika	1.6m
Red	1.8m

There is no denying that large expanses of tree tubes can be ugly, and for this reason many foresters and others responsible for amenity areas opt for fencing as a more acceptable alternative. They can also become redundant when a tree trunk begins to reach and then exceed the diameter of the tube, necessitating its removal.

Tree tubes in use to protect new forestry planting at the establishment stage.

Although many are now marketed as being biodegradable, many of the older types are not and will remain in the landscape for a long time if efforts are not made to remove and destroy them when they reach the end of their usefulness.

The major downsides of individual tree guards are the expense and the labour involved in placing them. The guards also need to be securely supported by stakes – more time-consuming labour – as they can blow over in high winds, and larger deer species will often deliberately push them over.

Once trees are established and too large for tubes to be useful, particularly valuable specimens may still need protecting and fencing then becomes unavoidable. Wrought iron or similar barriers may be appropriate to protect individual trees or small spinneys if a more decorative approach is desired. These are frequently seen in parks where deer, cattle or sheep may be present, and a number of manufacturers offer them from stock.

← Tree tubes offer not just protection but also a beneficial microclimate to encourage the growing plant.

← The use of wood or wrought iron guards, more decorative than tree tubes, may be appropriate in public places.

Hedging

An alternative to fencing is a suitably dense hedging plant. It may not be quite as efficient but can certainly be easier on the eye and can be combined with a simple low fence to enhance its effect. Obviously the wider it is the better, to deter deer from jumping it. Some people use climbing plants to break up the visual impact of a fence; additionally, deer are reluctant to cross obstacles when they cannot see what lies on the other side so it should be as dense as possible. Evergreen plants are the most effective; the likes of beech hedges, whilst undeniably attractive, become bare and less effective once the leaves have fallen in the autumn.

Rose hedges are occasionally cited as being effective, if allowed to grow thickly enough. Impenetrable vegetation in the form of the rambling rose, *Rosa multiflora*, is sometimes used in the USA apparently to create deer-proof barriers on some highways, although this species could become invasive if allowed to grow unchecked. More decorative rose varieties are more likely to have a place in the average garden.[3]

The 'dead hedge'

⬇ *A dead hedge under construction. Vertical poles are used to hold cut branches and other waste material in position.*

One very ancient form of hedging, dating back to at least the sixth century, is the so-called dead hedge, which was once an important means for the penning of livestock. As its name implies, no living plants are involved in the construction; instead it is built up using branches, foliage and other cut materials which are a by-product of forestry or other clearance activities. This not only finds a use for

Photograph David Lock

waste materials, but can provide an excellent habitat for birds, small animals and insects.

While cheap, environmentally friendly and easier on the eye than some other options, the dead hedge has a number of drawbacks. Only relatively small areas can be protected given the labour and materials involved, and deer (especially the smaller species) will probably find a way through them in time. Nevertheless, this type of hedge is still popular where a natural landscape is desired and especially in organic gardening. Dead hedges provide a very low carbon footprint, being constructed on site and avoiding any need for burning or transport for disposal. They are also a very effective way of deterring unauthorised persons from wandering into sensitive areas from public footpaths and the like and in addition can provide an effective windbreak.

To construct a simple dead hedge, material can simply be piled upon itself. Better yet, though, parallel hurdles can be constructed using stout uprights (slow rotting wood such as alder is most suitable) with a few horizontal lengths of more flexible growth such as hazel or ash woven between them to contain the other material which is built up inside. The hurdled approach takes up less space and is probably easier to build into a form that more effectively excludes deer, but the piled-up type can be made wide enough to deter larger deer from jumping it, and allows regenerating plants to grow up through it. As the hedge rots down it can be built up with fresh material, and the sunny side can even be planted up to create a living hedge.

The ha-ha

Most ways of excluding deer are based on constructions above ground level, but there is one notable exception. Visitors to deer parks may notice an unusual feature - a sunken ditch, virtually invisible from the house and gardens of the stately home, but which prevents the deer, or other animals, in the park beyond from accessing the immediate area of the house and gardens. This is known as a ha-ha, supposedly so-called because of the exclamations of surprise from visitors who suddenly notice it! The main point of the ha-ha is to provide an uninterrupted view from the main house onto the landscape beyond it. Such constructions usually feature a gently sloping incline on the livestock side, with a sheer wall (usually constructed of brick or stone) of sufficient height to prevent the animals from venturing beyond it.

Designed as a way to avoid the need for unsightly walls or fences, from the deer park owner's point of view, the ha-ha had an added advantage: wild deer could jump in but not out, thus replenishing stock in times when venison was often relied upon as a source of meat.

The ha-ha has fallen out of favour today, and they are of course relatively expensive to construct. If you do decide to go down this route, however, do bear public safety in mind. There is considerable potential for injuries to people unexpectedly coming across them, especially in the dark. In recent years there have been substantial personal injury claims awarded in favour of guests at stately-home events who have accidentally fallen into them – so the lighting, fencing and warning signs needed to prevent such claims being successful may negate the visual appeal of the ha-ha in modern times.

3 Coles, C. (1997), *Gardens and Deer*, Shrewsbury, Swan Hill Press

↑ *A typical ha-ha viewed from the livestock side...*

... and from ground level.

Passive Approaches: Deterrents and Repellents

Deterrents

There are a number of ways of persuading deer that they would rather be elsewhere, but their effectiveness may vary according to the species involved and local circumstances. What works in one place may often be found to be virtually useless in another, and what may be feasible in a small garden might be totally impractical when one is faced with large expanses of farm or woodland to protect.

Deterring deer from entering a given area can be difficult if large physical obstacles such as high fences are not possible. For smaller spaces, such as gardens or nurseries, you might consider a number of mechanical options for protecting specific areas. A cheap portable radio, set to a talk (rather than music) station and wrapped in a plastic bag to keep it dry, will often deter deer. One device is marketed which is connected to a hosepipe and activated by motion; it sprays a brief but powerful burst of water accompanied by a clattering noise. Deer, like any wild

➡ *There are a number of sonic devices on the market that claim to deter deer and other animals. This one offers adjustable sensitivity and noise frequencies.*

animal, hate anything unexpected. You could also look at motion-activated sonic devices, which can be effective under the right circumstances although users have reported mixed results. What is an unpleasant sound to one species is often inoffensive to another, and the ones which have frequencies that can be changed to suit the targeted species tend to work best: one fixed-frequency model trialled proved to have no effect whatsoever on deer but traumatised the family cat. There are plenty more deterrents on the market; a visit to your local garden centre or an internet search may suggest solutions which are worth trying to see if they will work for you.

A free-roaming dog will keep deer away if this is practical, but do consider neighbours who might not appreciate nightly barking. The same applies to any mechanical devices which produce noises audible to the human ear.

Deterring deer from entering larger areas is much more difficult. Farmers have long used scarecrows, but these are quickly ignored if they are not changed or moved regularly. Gas guns, initially effective, may also become less so with time as the deer come to realise that they are not a threat; the same applies to sirens or streamers. On more than one occasion I have watched families of roe, bedded in close proximity to gas guns, showing no reaction whatsoever when they went off.

Whatever you decide to try, it is important to ring the changes. Deer are adaptable and will soon come to realise if a perceived threat is in fact harmless: they will learn to ignore it.

Repellents

Deer repellents rely very heavily on the animals' keen sense of smell; they tend to be classified either as barriers, producing a sort of chemical fence that the deer are reluctant to cross, or as feeding repellents that are applied to the plants themselves that one wishes to protect.

You could try any number of simple deterrents in the garden or smaller area, although as ever success seems to vary according to the user. Lion dung is frequently quoted as effective. If you don't know a friendly zookeeper, pellets containing essence of lion dung are commercially available. I know of one local lady who sends her husband out every night to 'water' the vegetable patch and swears that the deer never trouble her plants, although doubtless a fall of rain would quickly dilute the effect.

Others suggest small bundles of human hair, contained in old tights, strung around the garden perimeter (these need to be replaced regularly as the human scent dissipates after a while). Moth-balls, on the other hand, are frequently recommended but have generally been found to be ineffective. Scented soap on the other hand can be hung off fences and branches; it seems to work better, and has the advantage of not having to be replaced as regularly. By all means try such methods – they may work for you.

Commercially produced chemical repellents for application on a larger scale are more restricted, and regulated in the UK by the Control of Pesticides Regulations 1986 and the Control of Pesticides (Amendment) Regulations 1997 (rather surprisingly, the use of lion dung and human hair also fall under these regulations). However, studies suggest that the chemicals licensed for use in the UK can be prohibitively expensive, have short useful lives after being applied, or degrade quickly and are only partially effective. One repellent, an Austrian preparation called Wobra, has been found to be very effective against bark stripping but is sadly not currently approved for use in this country.[4]

Diesel repels deer fairly well, and lends itself to creating an effective barrier across larger areas. Strips of rags dampened with it and suspended as a fence some four feet off the ground provide a relatively cheap and straightforward option, which easily enables a home-produced approach. Creosote is an alternative, although now not so readily available to any but professional users such as farmers, and does not seem to be as effective as diesel in any case.

4 Putman, R. anc Langbein, J. (2003), *The Deer Manager's Companion*, Shrewsbury, Swan Hill

There are various commercially produced repellents on the market which can be painted onto a tree or shrub, mainly to prevent bark stripping, but generally these can only be applied during dormant periods. Once new growth occurs, the new shoots are unprotected and vulnerable. Very few chemical repellents are suitable for application to growing foliage. The expense involved, though, limits the size of an area that might be realistically protected in any case.

The Forestry Commission produce a number of advice notes and best practice guidance documents, many of which are downloadable from their website.[5]

Haymaking casualties

Every year there are casualties among young deer during silage cutting and haymaking operations. The problem is especially associated with roe, the majority of which are born between mid-April and May. Kids, left alone for long periods while their mothers leave to feed, will instinctively freeze if danger threatens and roe kids do not fully grow out of this habit until they are about a month old. During this time they are particularly vulnerable to agricultural machinery.

Modern agricultural practices, which involve intensive harvesting rather than the old fashioned 'circular' cutting, allow the deer less opportunity to decamp, and modern machinery is larger and faster moving than that used in the past. Many methods have been tried to reduce casualties, including chemical preparations or flushing bars fitted to vehicles, but simply making the area unattractive to deer through human presence is ultimately more sensible than trying to find immobile deer fawns at the time of the cut. Hazel wands with white flags attached, placed around the field to be cut at intervals of some 80 metres, has been found to be an effective method by some, as has the regular use of reliable dogs, flashing lights or noise-makers in the days preceding cutting. The mother will then simply choose a different place to give birth, or decide to move her fawn to somewhere she considers to be quieter and safer.

5 Trout, R. and Brunt, A. (2014), *Protection of Trees from Mammal Damage*, Forest Research Best Practice Guidance No 12; available from www.forestry.gov.uk

← *Most garden plants are vulnerable to hungry deer; these runner beans have been browsed by roe.*

Passive Approaches: Deer-resistant Plants for Gardens

The list of cultivated plant species vulnerable to deer damage is long and sadly includes many popular ones such as clematis, crocus, fuchsia, lupin, pansy and rose. There are very few plants which can be described as truly deer-resistant, as they will try anything new, and the list of those that are truly unpalatable to deer is rather short. Furthermore, gardeners who restrict themselves to just those species will find only limited opportunities to improve the diversity and appeal of their garden.

The British Deer Society[6] suggests the following plants as being particularly deer resistant:

Camellia	*Poppy*	*Hosta*
Iris	*Fuchsia*	*Sedum*
Cistus	*Primula*	*Hydrangea*
Lavender	*Hellebore*	
Crocus (some species)	*Rhododendron*	

All is not lost, though. By allowing natural food plants to become established, the attention of deer can be diverted from prize specimens. The following are suggested as suitable diversionary plants: brambles, rosebay willowherb, rowan (mountain ash), dandelion, campion, hoary cinquefoil, knotweed, sweet lupin, redleg, ribwort and yarrow. If such alternatives are provided, the following have shown themselves as capable of avoiding excessive damage:

Agapanthus	*Lavender*	*Choisya ternata*
Cornus sanguina	*Rosa rugosa*	*Honeysuckle*
Juniper	*Birch*	*Pampas grass*
Potentilla fruticosa	*Foxglove*	*Viburnum (deciduous)*
Alder	*Lupin*	*Chrysanthemum*
Cotinus coggygria	*Shallon*	*maximum*
Kerria japonica	*Box*	*Hippophae rhamnoides*
Ribes spp.	*Gaultheria shallon*	*Philadelphus*
Aquilegia	*Magnolia*	*Vinca spp.*
Daphne spp.	*Snowberry*	*Cistus*
Kniphofia	*Buddleia davidii*	*Hydrangea*
Robinia pseudoacacia	*Gooseberry*	*Phormium tenax*
Azalea (deciduous)	*Mahonia spp.*	*Weigela*
Delphinium	*Spiraea japonica*	*Clematis spp.*
Lonicera nitida	*Chaenomeles*	*Jasmine*
Romneya coulteri	*Hellebore*	*Pine*
Berberis spp.	*Narcissus*	*Yucca*
Forsythia	*Sweetbay*	

The Royal Horticultural Society provides similar listings on their web page, which is updated regularly with recommendations from members.[7]

6 The British Deer Society (2015), *Deer in Gardens*; available from www.bds.org.uk/index.php/advice-education/deer-in-gardens (Accessed 12 February 2018)

7 The Royal Horticultural Society (2018), *Deer resistant plants*; available from www.rhs.org.uk/advice/profile?PID=185 (Accessed 22 March 2018)

Passive Approaches:
Planting and Feeding

Decoy crops

If gardeners can successfully divert the attentions of deer with suitable alternative fodder, where valuable woodland or other areas are receiving unwelcome damage it may sometimes be cost-efficient to offer deer attractive alternatives there too. In one case a chief forester regularly planted willow rods that would eventually sprout and provide an alternative food source to the spruce crop that he was trying to protect, willow being very attractive to roe. This proved highly successful at diverting their attention from the vulnerable spruce whilst it established itself. Willow is particularly quick to grow from rods but other species, such as ash and rowan, can also be effective.

Many managers of land with game-shooting interests, where game cover is deliberately provided, or indeed farmers who grow fields of specific crops, will recognise how deer are attracted to deliberately planted areas during the different seasons. Decoy crops may well be appropriate in some places, offered as sacrificial alternatives to more valuable ones elsewhere; where culling has to take place, these areas can also provide suitable places for this to be carried out. Such crops can be especially valuable during the late autumn and winter when more natural food sources may be scarce. Root crops such as turnip, swede and radish varieties, as well as kale and rape, are all good alternatives. Legumes such as peas and beans are very popular earlier on. Brassicas, including cabbage and broccoli, are readily taken too. During the late winter, freezing temperatures will also raise the sugar content of leaves and increase their attractiveness to the deer.

Where fraying damage is an issue, sacrificial saplings can be planted to distract deer, particularly roe, from more valuable plantings. Roe typically select their fraying stocks in prominent parts of their territories and return to them time after time to fray and scent-mark them. This can often be along the edges of commercial plantations where thoughtfully provided alternative saplings may well be selected by them.

Deer lawns

You often hear cleared areas in woodland, specifically planted to attract deer, described as 'deer lawns', even though the term tends to give something of a false impression. The aim is not to produce a flat, open area consisting entirely of grass - deer seldom feel comfortable in such places - but rather one with scattered light growth and occasional trees as well as suitable forage, and heavier cover available close by to allow a safe retreat if necessary. Good soil and sunlight to facilitate growth are important, and a lack of disturbance is essential. Deer will naturally gravitate towards such places to feed and rest.

A suitable lawn can be simply created by some basic scrub clearance and an occasional thinning to stop it becoming overgrown, but more deliberate efforts can be rewarding. This has long been recognised in many other countries but the concept is becoming increasingly popular in Britain, and seed suppliers now offer purpose-made mixes for deer. One commercially available mixture puts the emphasis on meadow fescue, comprising about a quarter of it, with ryegrasses and meadow grass also being prominent. Other plant species featured are sainfoin, lucerne, white and red clovers, chicory, foxtail and fenugreek. The seed is sown at a rate of some 37 kilograms to the hectare (half-hectare plots are considered a suitable size for individual lawns), and the suppliers recommend it as supplying the needs of most British deer species.

Brambles are a very attractive plant as far as deer are concerned, offering a combination of food, shelter and protection.

A diverse selection of plant growth is the key, and there are many others that are attractive to deer. Variety ensures that the deer receive the balanced diet that they seek, and that they continue to revisit the plot regularly throughout all of the seasons. Planting just one type of seed in basic food plots encourages only occasional use by the deer. It goes without saying, of course, that a suitably established deer lawn is beneficial to many other kinds of wildlife as well.

Supplementary feeding

In some countries it is traditional to provide supplementary feed to deer, particularly during the winter months, both to sustain them and to boost body condition, breeding success and antler growth. Sometimes the aim is to support a higher deer population than would normally be possible for the ground in question; this is mostly the case where the land is under heavy hunting pressure. Fallow and red deer seem to take most readily to this 'hand' feeding, roe less so. In the UK, however, the more usually accepted approach is to try to maintain deer numbers in balance with their environment without any reliance on artificial feeding. Intensive natural feeding in the months leading up to winter usually allows the deer to lay down sufficient fat reserves to fall back upon when food is scarce.

A continental deer-feeding station. Such supplementary feeding is not the norm in the UK.

An artificial feed mixture of cake and grain.

There are times, however, when prolonged periods of snow and ice mean that fat reserves can be quickly used up and it is then that the deer start to suffer. At such times it often seems that providing artificial feed is the answer. Sadly, though, changes to a deer's diet have to be made slowly as their stomachs - or more accurately, the specialised protozoan flora that exist in the rumen to ensure efficient digestion - cannot instantly adapt to the new diet. This process can take as much as two weeks. As a result, it is possible for a starving deer fed something that it is not used to, such as hay, to die of malnutrition despite having a full stomach. Emergency feeding, regrettably, can often come too late to have any positive effect. Continental experience suggests that artificial feeding should commence as early as October to allow animals to get used to it and build up fat reserves before the arrival of hard weather. Where they do not take readily to the strange offerings, more familiar ones such as ivy, Douglas fir or spruce can be hung on hoppers to make the feed more acceptable.

Photo: Owen Beardsmore

➥ *A roebuck visits a woodland feeder offering a wheat and feed bean mixture.*

Any foodstuffs offered need to be cheap, readily available and above all palatable. Concentrates such as deer cobs are easy to acquire, while mixtures based on constituents such as maize, sesame seed, copra cake, oats, wheat or peas are also popular. Supplements to them, such as apples, chopped root crops or hay (preferably cut from old pastures with a lower proportion of grass and a higher one of flowers and buds), can assist with rumination, but rotting vegetables and other discards are unlikely to be attractive. 'Forest hay', consisting of cut and dried leafy branches, or ivy provide a more natural diet.

There are some drawbacks to supplementary feeding. Large concentrations of deer may encourage build-ups of parasitic infection, and there is the inevitable risk of an increase in deer-related damage in the immediate area. With many species, the more dominant animals may take over the hoppers and deny others access; placing several hoppers well spread out at each station will help to reduce quarrelling between animals. Beware, too, that a feeding station may itself attract poachers if the location becomes known, so these need to be placed well out of the public eye.

Artificial feeding should never be undertaken lightly, though it can form part of an overall management plan and exceptional circumstances may justify it. If the deer population is maintained in tune with its environment, however, it is unlikely that it will be necessary or desirable under normal circumstances.

It is worth adding that some stalkers provide feed to bait deer into known areas for the purpose of shooting. This is only likely to be consistently successful with the smaller, more solitary species; herding animals will quickly learn to avoid the area. Field beans offered from specially designed dispensing hoppers seem to be particularly successful, and chopped carrots or apples may be popular – although the readiness to accept such offerings may vary from place to place. The practice is certainly legal (and many deer stalkers deliberately site their high seats to cover pheasant feed hoppers for the same reason) and many consider the practice to be ethically justifiable if local conditions make it difficult to locate deer where cull targets have to be met.

↑ Supplementary feeding is only effective if deer have had time to accustom their digestive systems to new food sources. Here, a herd of park sika enjoy their regular winter supplement.

Mineral licks

Another long-established practice on the continent is to provide mineral licks for wild deer and this idea too is increasing in popularity in the UK. That deer need

➡ A mineral block in place on a rock. The deer generally take the dissolved salts as rainfall causes them to seep out towards the ground.

trace elements to develop their physical condition and antlers is beyond question but they usually obtain these elements through regular natural foraging and should not normally need artificial supplements.

Mineral licks are credited with encouraging higher body weights among adult animals and subsequently birth rates. There have certainly been recorded instances of male deer that regularly visit them producing antlers of a higher than normal standard for the local population. The best way to establish a lick is to fix the mineral block to the top of a post or tree stump; as the block is dissolved by the rain it will run down the post onto the ground where it is more readily taken. The addition of an attractant, such as aniseed, may help to draw the deer to them.

Some deer species seem to take more easily to licks than others. In the UK, fallow tend to visit them most readily, whereas roe seem much less likely to touch them and the only muntjac I have ever seen using one was in captivity. Generally speaking, though, wild deer frequently appear reluctant to use them even in poorer habitats. On the continent, deer of all species are rather more likely to use licks but this could simply be due to conditioning to the habit from a longer history of them being available.

By all means try a mineral lick or two on your ground. If vital trace elements are deficient in the area for whatever reason they may well prove beneficial and they can provide useful opportunities for observation. Beware, though, that (just as with feeding stations) the deer will quickly learn to avoid them if culling or other disturbance takes place at the site.

Feeding deer in gardens etc

Some people, rather than wanting to exclude them, enjoy seeing deer in their gardens and deliberately leave food out for them. There is really no need to feed visiting deer, as they can usually find all that they need naturally and may actually ignore any offerings completely. Some householders still decide to provide their visitors with an occasional treat, and in such cases deer may accept any number of raw vegetables. Brassicas, such as cabbage or sprouts, and field and green beans can be very popular. Chopped carrots or potatoes can also be attractive, and there have been reports of deer enjoying peanuts from bird feeders. As they will take wild fruits and fungi according to the season, so mushrooms, shelled nuts and chopped apples or other fruit might also tempt them. Food should be left on clean ground and always cleared away quickly if it starts to rot.

It is important to remember that, as deer do not have upper incisor teeth, they may have difficulty dealing with larger hard items such as whole root crops or apples, so such offerings do need to be chopped into slices or pieces. Some sources suggest that you can leave out bread or table leftovers but this is not advisable as any processed or cooked food items should be avoided. Anyone putting food out should always be mindful of their neighbours, who might have different feelings about encouraging deer into the vicinity of their properties.

Deer are great opportunists and some learn quickly to overcome their natural fear of man in return for an easy mouthful or two. Regrettably, feeding deer and other wild animals at roadsides and other public places only encourages them to hang around there hoping for a hand-out, and this greatly increases the risk of a collision with a vehicle. Another important consideration is that while it may be tempting to feed deer from the hand, great care needs to be taken. Wild and even park deer can be nervous animals and a sudden fright might cause them to lash

⬇ *A deer's hooves are sharp and readily used for both aggressive and defensive purposes. Here two sparring park red deer, their sensitive growing antlers encased in velvet, 'box' with their forefeet.*

out. A deer's hooves are very sharp and are frequently used in self-defence, while it only takes a slight movement of a stag's head for its antlers to cause a serious injury to someone standing close to it – even if this is not intentional.

Deer and roads

Unfortunately, as traffic on our road networks has grown, so has the number of vehicle collisions involving a deer population that is also increasing. There is no legal requirement to report road traffic accidents involving deer so we have no way of knowing exactly how many occur; but the Highways Agency estimated as many as 74,000 in 2017 alone, including between 400 and 700 human injuries and 20 deaths. Amongst the deer themselves, the number will be much higher. Although many are killed outright, more yet will move off with injuries which either mend or cause their later demise.

The highest risk tends to come during the October to December period, a time when deer are moving more for the rutting season. Otherwise deer movement is higher at dawn and dusk so, perhaps inevitably, the likeliest time of day for a collision will be between sunset and midnight, and also shortly before and after sunrise. When these times coincide seasonally with busy commuting periods during the autumn and spring the risk increases further. Many accidents occur when a driver swerves to avoid hitting a deer, subsequently colliding with other objects or road users. Driver awareness is certainly a major problem; many people see a deer cross the road but will not slow down, failing to realise that more may follow. Drivers can also become so used to warning signs, even when they are prominently sited, that they come to ignore them.

↓ *Red stags crossing a road on Exmoor.*

Photograph: Jochen Langbein

Risks can be mitigated by some fairly simple measures. Most accidents seem to occur in heavily wooded areas or where dense hedges are sited directly alongside roads. While appropriate fencing remains the main proven method of reducing accidents, it can be highly expensive to cover large areas and regular maintenance is needed; accordingly, efforts tend to be restricted to identified high-risk sites. Fences are likely, of course, to redirect deer to crossing points elsewhere. There are options, though. By allowing a clear verge of grass or low vegetation, which the deer can easily see over between thicker cover and the road, the chances of them running blindly into traffic are much reduced. Studies in the USA have concluded that the optimum width for this open strip needs to be between 20 and 30 metres.[8] There has also been some success with scent barriers, which discourage the animals from accessing the verge at particularly hazardous crossing sites, instead diverting them to other more suitable places where they are more readily visible to motorists.

Commercially marketed sonic devices designed to be attached to vehicles have long been available, but there is little evidence to suggest that they are especially effective.

Much experimental work has been conducted with these and also roadside reflectors. Where traffic volume is low, devices that reflect vehicle headlights may have value in deterring deer from crossing until there is a suitable gap in the vehicle flow, but in high volume areas this is less likely to be effective. Work continues to be conducted in this area and results so far are encouraging.

The problem does not, of course, just affect this country and many others are undertaking some very novel and innovative research. In Japan, for instance, considerable train delays were being caused by collisions with sika deer which were attracted to the tracks, apparently to lick iron filings from them. Now, sound systems fitted to trains and used to broadcast animal noises (such as deer alarm calls linked to dogs barking) appear to have reduced animal sightings in the vicinity of tracks by as much as 40 percent.[9] This move followed earlier, less successful, attempts to deter the deer with fencing, which the deer simply walked around, or spraying lion dung onto the tracks.

The use of under and overpasses has been extensively studied and proven to be effective as points for wildlife to cross busy roads. Pioneering work in Germany has determined that there is a direct relationship between the opening and lengths of underpasses is critical if animals are to use them, and even then it can take considerable amounts of time (six months for roe, and as many as three years for other deer species) before the deer overcome their wariness of using such structures.[10] Deer do not readily take to using tunnels, and the overpass can be more readily accepted, but such structures need to be wide enough to accommodate them and planted appropriately.[11] As a result they are not cheap. While the cost of a small culvert crossing constructed in the USA might be around US$10,000, two full-scale wildlife overpasses constructed over the Trans-Canada Highway in 1997 cost C$1,851,000 or just over £1 million each.[12] A US study of underpasses concluded that they only became cost-effective if they prevented between 2.6 and 9.2 deer/vehicle collisions per year, depending on the cost of the underpass.[13] Given the comparably much higher costs of overpasses, this number must be considerably greater to justify the latter in financial terms.

8 Putman, R. and Langbein, J. (2003), *The Deer Manager's Companion*, Shrewsbury, Swan Hill p66

9 *The Times* (London 2018), report: *Snorting, barking train scares sika deer from tracks in Japan*, 17 January 2018

10 Olbrich, P. (1984), *Study of the effectiveness of game warning reflectors and the suitability of game passages*, Zeitschrift für Jagdwissenschaft, 30, 101–116.

11 Simpson, N. O. et al (2016), *Overpasses and underpasses: Effectiveness of crossing structures for migratory ungulates*, The Journal of Wildlife Management, Volume 80 Issue 8

12 Bissonette, J. A. et al (2007) *Evaluation of the Use and Effectiveness of Wildlife Crossings*, NCHRP Report 615; available from http://wildlifeandroads.org (Accessed 20 February 2018)

13 Donaldson, B. M. (2005), *The Use of Highway Underpasses by Large Mammals in Virginia and Factors Influencing their Effectiveness*, Virginia Transportation Research Council; available from www.virginiadot.org/vtrc/main/online_reports/pdf/06-r2.pdf (Accessed 19 February 2018)

Photograph: Jochen Langbein

⬆ *A wildlife overpass or 'green bridge' in Germany.*

⬇ *Shotgun (left) and rifle cartridges. The rifle offers accuracy at longer ranges with bullets designed to perform predictably and deliver maximum terminal energy. The shotgun is only effective at short ranges and the law permits its use only for killing uninjured deer under very specific circumstances. Under normal conditions it is best reserved for humane despatch.*

The hard fact remains that, as deer numbers increase, so will incidents involving vehicles, and in the long term the only way to mitigate this risk is to reduce the overall deer population. In reality, less deer will mean less accidents.

Active Approaches

Sometimes a passive approach to deer control is just not possible and it becomes necessary to reduce the overall numbers. In the UK the law is very specific about how this may be done: essentially there are two tools for the task which we will examine in more detail over the coming chapters. The full legal requirements for firearms used to kill or take wild deer are addressed in Chapter 10.

The rifle

For the stalking and killing of uninjured wild deer, the equipment employed must enable the placement of a single, lethal shot with as small a margin for error as possible. The full-bore rifle is without doubt the best tool for this. It has a long, effective range in practised hands, which means that animals can be studied at leisure to allow their age, sex and condition to be assessed before the decision to shoot is taken. A correctly shot deer will therefore be culled instantly and painlessly.

Modern developments have made the rifle capable of excellent levels of accuracy at working ranges of beyond 150 metres, enhanced by telescopic sights which allow shot placement to be even more precise. Rifles are also easily fitted with sound moderators which reduce the sound signature, resulting in less disturbance to the general environment while reducing recoil (and thus enhancing accuracy) as well as protecting the shooter's hearing.

It follows that the rifle is unreservedly the best tool for the humane control of deer.

The shotgun

The shotgun, in the right hands, at the right range and with the right cartridge load, is certainly capable of humanely killing deer but there are too many variables at work here. Even using large shot, it has a relatively short realistic range of no more than 40 metres or so, and lacks the extended reach and precision of the rifle. Furthermore, at close ranges the deer will probably be aware of the shooter and be on the move already so there is a reduced chance of a perfectly placed shot. Very importantly, the chances of having sufficient time to identify and assess the deer in question will be much reduced when it is fleeing, so selective shooting is less likely and there is an increased chance that out of season animals might be accidentally taken. The law is very prescriptive in the UK about the limited circumstances under which a shotgun may be used for the killing of healthy wild deer, as noted above; even when permitted, it is not a suitable tool for mainstream culling activities.

The shotgun is, however, very well suited for one aspect of deer management: the humane despatch of badly injured animals. Small calibre shotguns discharging small shot pellets, which are very effective at close range with a rapid loss of energy thereafter, are especially suitable for this task.

Handguns and captive bolt devices

The possession and use of handguns is very heavily restricted in the UK, and users must demonstrate an exceptionally good reason to hold one. Being short-barrelled and supported only by the shooter's hand or hands, they are relatively inaccurate even at short ranges when compared to a rifle. Most do not deliver sufficient energy to meet UK legal requirements for shooting deer in the first place. While there is sometimes justification for their use for the humane despatch of injured animals, there can be none for shooting at uninjured ones with them. Furthermore, if a deer is inadvertently wounded, the hunter should dispatch it as soon as possible with another shot from the firearm already being carried and there is no case for the additional carriage of a handgun for this purpose.

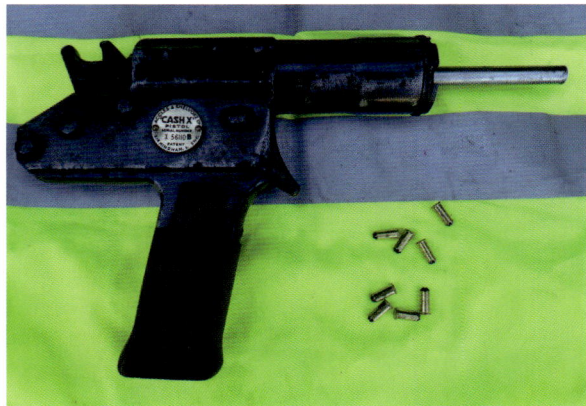

↑ *A captive bolt device, shown with the piston fully extended and the .22 blank cartridges that power it.*

Captive bolt devices are instruments used in animal slaughter that project a restrained piston for a short distance, normally no more than a few inches, beyond the end of the barrel. They are sometimes referred to as 'humane killers', a term which is not strictly accurate as they are actually intended to render an animal insensible prior to actual despatch. A captive bolt device requires skilled handling although, unlike a handgun firing a free bullet, no certificate is required for their purchase, possession or use. A valid slaughter licence is, however, generally required for their use on livestock in all but emergency or casualty situations. Like the handgun, they have no place in wild deer management beyond humane animal despatch.

Bows and arrows...

There is increasing pressure in some quarters to permit the hunting of deer with bows or crossbows, currently illegal in Britain. Proponents are especially quick to compare the reduced lethal range of an arrow when compared to that of a full-bore bullet, which might be capable of killing at well over two miles if fired carelessly. This, they say, is a perfect solution to problems with deer encountered in built-up areas.

While there is little doubt that a bow in the hands of a skilled user will kill a deer quickly and humanely, as has been demonstrated in other countries where bow hunting is legal, the emphasis must be on the word *skilled*. It is relatively easy to learn to use a rifle accurately, whereas a bow requires constant practice as well as much greater fieldcraft skills if an approach is to be made to within a suitable range for the shot – which, like the shotgun, will be relatively close.

There are also issues of availability and licensing; the legalisation of bow hunting would have not only potentially serious deer welfare implications, but probably also result in increased instances of poaching. Given these factors it is difficult to imagine circumstances where the legalisation of bows for killing deer can be justified in the UK.

A non-lethal alternative?

The use of contraception is regularly raised as a humane alternative to killing deer where numbers have become, or are in danger of becoming, excessive. The idea is certainly attractive in theory so it is worth examining the idea in more detail.

Most people think of contraceptive pills or injections in human terms, where treatments are based on the use of steroidal ovarian hormones (oestrogen and progesterone) to suppress ovulation. Similar treatments used on wild animals, however, can have adverse effects on the environment, the natural food chain and the natural behaviour of the deer themselves. Steroid hormones like the ones used in human contraceptives would adversely affect antler growth and development. As a result, the alternative developed for use with animals works differently, using the animal's immune system to render it temporarily infertile, hence the term 'immunocontraception'. This method has already been used successfully to control numbers of horses and deer in North America, and elephants in Africa, with very few side effects noted.[14]

Immunocontraception works by injecting the animal with a protein very similar to one of the natural proteins essential for reproduction. This might be the protein in the jelly that covers the unfertilised egg, or a protein released from the pituitary gland. The animal then makes antibodies to the protein in question and these antibodies disrupt the complicated process of ovulation or fertilisation. In this way pregnancy is prevented, although the animal will otherwise continue to behave perfectly naturally.

Treatment needs to be administered regularly to be effective and it is here that problems arise. The trials on American white-tailed deer, whilst successful, were mainly conducted on captive populations where each animal could be identified individually and easily captured for an injection to be given. Wild deer obviously cannot be handled and so have to be darted to avoid causing panic and the

14 Rutberg, A. and Naugle, R. (2008), *Population-level effects of immunocontraception in white-tailed deer (Odocoileus virginianus), Wildlife Research*, 2008, 35, 494–501

possibility of animals damaging themselves. The darting systems currently available are generally only effective at impractically short ranges, and there is no way of confirming that a vaccine has been successfully delivered. Of course the animals also have to be recognisable as individuals, to ensure that the right ones are given the appropriate treatment.

Where immunocontraception has been used on populations of deer, it has taken several years for the population growth to slow down and wildlife biologists currently advise that population reduction can only be achieved by culling or live removal. Research continues and there are hopes that a practical solution may yet become available. Regrettably, at the time of writing (in 2019), this prospect still seems some way off and one wonders if the expense involved will ever be practical or justifiable in the majority of cases. ■

4 Active Management: Planning and Preparation

Before the active management of deer can even begin there is much that needs to be considered: launching yourself blindly into a situation seldom achieves the best results.

Woodland design for deer management

All too often the difficulties of managing deer populations are not taken into account when new woodland plantings are in the design stage, causing later difficulties when tree growth has occurred and substantial deer numbers and their effects need to be minimised.

Wherever possible, plantations should not be too large; when they are, deer can often seem to disappear into them and then exist without being seen easily. Square blocks should be avoided. Heavier levels of browsing are to be expected around the edges of plantations, so those with long sides are more liable to suffer damage – although irregular edges do allow for easier approach by a stalker, while encouraging the deer to feel more secure and thus to be more likely to venture out of cover. It also follows that the less holding cover that there is around the edges of plantations, the more likely it is that damage levels there will be greatly reduced.

If clearings are left within the new areas of planting, ground cover will be encouraged, itself providing suitable alternative food for the deer while attracting them to both feed and lie up. Such clearings also enable the deer to be more readily seen within a wooded area and offer the perfect position for high seats, so they need to be designed to allow for access, observation and sensible shooting ranges if necessary. They should not be entirely empty of trees and larger bushes, as these will add to a sense of security for the animals using them.

Where access roads are needed to allow for the extraction of carcases, especially where the larger deer species are concerned, these can be combined with a firebreak function. Allowing vegetation to grow up along the sides of such rides provides browse which will encourage the deer out of cover. Rides and other tracks should not be straight; frequent corners permit easier stalking and also avoid them becoming

⬇ *Clearings with discreet access points allow them to be approached and overlooked more easily.*

➡ *Open verges at the side of woodland roads and rides permit easier visibility of moving deer. This high seat has good lines of sight in three directions from a track junction.*

wind tunnels that can deter deer presence. The footprints of power lines, while inevitably straight, provide similar opportunities and can be seeded deliberately to provide 'deer lawns'. Watercourses can also be attractive, especially if left unplanted for 20 to 30 metres on either side to provide wildlife habitats. These, too, provide suitable stalking routes in an area naturally attractive to deer.

Overhead cover adds to a feeling of safety and increases an area's attractiveness. Weeding should be undertaken sparingly along the sides of paths and rides as it provides alternative, and possibly more appealing, sources of forage. High weed growth, especially bramble or coppice, can also afford young trees a degree of protection in addition to its food value. Intense weeding along rows of young trees allows the deer not only easy access to them but also a natural path along which to feed. More localised clearance, limited to the base of each tree, does not provide such a path and limits the animals' access.

The best time to conduct new plantings is when adjacent cover is at a bare minimum, with little ground cover or after brashing (the removal of the lower branches of trees, usually conifers, up to a height of two metres) or thinning. This ensures that the general area is more exposed to view and the elements and as such rendered less attractive as far as the deer are concerned.

A few decoy trees of the right size and 'whip' can distract fraying activities, especially among roe, if correctly sited in likely visible positions for territorial marking - although very often it can be difficult to ascertain this from a human viewpoint. Oaks, willows, many of the poplars, lodgepole pine, Lawson's cypress and maples are considered to be especially attractive to deer. Otherwise, ornamental or trees unusual for the area will almost inevitably attract the attention of deer and may require special protection.

Building relationships

Deer evoke a wide variety of emotions among people, according to their profession or interests. The farmer or forester may see them simply as a threat to their livelihood, the ecologist may consider them either a vital element of overall diversity or a baleful influence which reduces the habitat available for other creatures, while the casual observer simply enjoys seeing them from time to time when jogging or walking the dog. Others may feel that deer must be preserved at all costs and will not countenance any talk of control at all. The deer manager must maintain a delicate balancing act between all of these viewpoints and somehow navigate a careful path through them.

Building good local relationships is an essential part of the job and the best approach is to be open, honest and talk to people. It is vital to recognise that other land-users have a right, and often a pressing need, to be there and not to assume that deer have a particular and special importance (although it is understandable that, once immersed in the subject, it can often feel that way). Gamekeepers in particular can be very protective of their woods, especially when there are young poults in the release pens or the shooting season is under way, and it pays to demonstrate that your activities are not disturbing them. In all cases, an occasional chat over a cup of coffee will help to deconflict land usage, build up a mutual understanding and enhance your position as an extra set of eyes and ears on the ground during the more unsociable hours. Even where it is not necessary, the occasional gift of venison often goes a long way.

You will also find that you start receiving reports of deer sightings and other matters of interest and will be regarded as an asset. In return, timely information of any unusual occurrences on the land in question is always appreciated by your new contacts.

Permission

Unless you are fortunate enough to own the land it is essential that you secure the appropriate authority to be on it, so the landowner is probably the most important person that you need to consider. Quite apart from determining the management objectives (which will be considered in Chapter 7), the landowner will very likely wish to determine the terms of your access to the ground and may also want to place some restrictions on access or activities.

Obtaining written permission, which can be produced if challenged, is highly advisable and indeed a legal requirement. It not only helps to avoid any potential for misunderstandings in the future but also enables a more professional approach which benefits all concerned. Although some situations may demand a more formal contract, under most circumstances a simple letter is usually sufficient. Following discussions, the deer manager is probably best placed to produce a suitably worded format for the landowner's signature. It should include such aspects as:

◗ Scope of activities (including target species)
◗ Access points
◗ Notification of access (if required)
◗ Authorised firearm types
◗ Infrastructure (such as building and maintaining high seats, and any associated pruning or felling))
◗ Risk assessment and responsibilities

⟩ Assistants/guests
⟩ Disposal of carcases and any funds generated from sales
⟩ Records
⟩ Reporting

An example of a simple letter of permission is at Appendix VI.

Neighbours

Liaison with neighbouring land-owning interests is definitely something that needs to take place before any active management commences, and is without doubt an area where problems can easily arise if prior agreement has not been sought. As remarked upon elsewhere, different landowners may have widely varying attitudes towards deer and it is important to find common ground - or at least to reach an acceptable consensus.

Even where the establishment of a formal Deer Management Group (see Chapter 8) is not possible or necessary, information can be shared, overall cull figures agreed across a travelling deer population's real range, or practical cooperative measures enabled where animals are pushed across boundaries by activities on one side of them. In many cases it may be possible to agree access actually to cull on a neighbour's property or along the boundaries, or erect high seats there to enable better lines of sight on regular movement routes.

At the very least, access arrangements can be made where approaches from one side are more difficult when it is necessary to erect structures such as high seats or recover carcases. Permission to cross boundaries to conduct the pursuit of inadvertently wounded animals or those that fall just over them is useful to hold in advance, and both saves time and avoids accusations of trespass or worse. This is especially helpful where there are roads in the vicinity with their attendant risk of deer/vehicle collisions. Wasting time contacting a landowner or farm manager to agree access can mean all the difference between a recovered animal and one that is lost.

Wherever there is any recognisable potential for future disagreements, it is always wise to keep a written record of any understandings reached, even if it is simply in the form of an email trail. This should take care to detail who actually owns any carcase of a legally killed deer that needs to be recovered from across a boundary.

⬇ Deer, wounded or otherwise, have no respect for boundaries set by man.

← *Time spent learning the ground and evaluating the deer on it will pay dividends when actual culling commences.*

Reconnaissance

As the old military adage points out, time spent on reconnaissance is seldom wasted. Time spent talking to landowners or those who work the land is invaluable, taps into local knowledge, and sets the tone for future relationships. It often gives you a good idea of where to start focusing your efforts. A large-scale map is extremely useful, so that any boundaries and sensitive areas (such as pheasant pens or conservation areas) can be marked down. Areas of high risk such as public footpaths and other places of access, roads and houses, should also be clearly identified. At the same time, it is helpful to enquire about regular activities, such as the feeding of farm stock, so that these areas can be avoided; regular liaison will ensure that such information is kept up to date and the deer manager informed of any future plans or changes to the routine.

With all this information available, a comprehensive survey of the ground under management can be carried out. This can identify suitable lines of approach, where shooting may not be safe because of the topography of the ground, or where high seats might be advisable. Take care to give some thought to carcase extraction and general vehicle movement; it is helpful at this stage to determine the landowner's views on where vehicle traffic is inappropriate and may cause damage to fields or other places where the going is not solid enough to support it. Many deer managers have found cause to be grateful for a telephone number to contact farm workers or others who live on site and are available and willing to assist in emergencies or with the extraction of larger carcases.

Time spent on the ground at this stage, armed with nothing more than a pair of binoculars and a notebook, will pay off in the long run. This will provide not just a picture of what deer are present, but what their movements are and what the sex/age balance is. Such knowledge can also be backed up by the deployment of trail cameras and taking the time to speak to other land-users. In this way the observer will also build up a wider knowledge of local circumstances and trends which might otherwise tend to be overlooked when focused on specific activities such as erecting high seats or stalking.

Risk assessments

Risk assessment is a subject that causes unnecessary alarm to the uninitiated, but really it is a very simple process that relies heavily on common sense. In essence all that needs to be done is to identify the risks, decide who might be harmed, work out the controls necessary and then keep records of any action taken where necessary. These assessments generally apply to work places, where they tend to be the responsibility of the employer, but anyone taking on deer management responsibilities on land which is not their own will probably find themselves required to produce one. In this way the landowner is fulfilling a duty of care towards those who work on his or her land.

As far as deer management operations are concerned, the process should start with the responsible person considering the conditions that they, and any assistants, will be working under. Much of this will be in the field, although if there is a deer larder or other infrastructure that needs to be looked at as well. The process should start with a comprehensive visit to the ground involved, a talk with the landowner and any other workers, and of course the management team themselves. Special attention should be paid to areas which present special hazards, and if a local 'accident book' exists, that can provide a useful record of any past problem areas. After careful consideration a list can be created, showing who might be harmed by the hazards and how. Once these have been identified it is a straightforward matter of devising suitable controls to manage the hazards, and any further measures that might be implemented. Any actions that need to be carried out in the future are also identified, and a time-frame set to implement them.

Once completed, the risk assessment should not be put away and forgotten but displayed in a public area such as a rest room or, perhaps more practically, provided to all concerned either in hard copy or electronically. It should be updated and reviewed regularly; annually is usually a sensible time period.

A sample risk assessment is at Appendix IV. While it is based on deer culling, it can easily be adapted for other activities such as infrastructure construction or census days, and local circumstances should be considered. A brainstorming session with others involved in the activities is often very helpful and can bring out any issues that may not have occurred to the person producing the assessment. Whilst it is easy to fill a great many pages for a risk assessment, in reality most people find that it is not necessary to identify more than a dozen or so relevant hazards.

If further advice is needed, the Health and Safety Executive provide an easy to understand guide which can be accessed on the internet.[1]

Insurance

Some thought needs to be given towards insurance. While it is true that landowners have a legal liability for activities which take place on their land, they will usually

1 Health and Safety Executive, *Risk – controlling the risk in the workplace*, available from http://www.hse.gov.uk/risk/controlling-risks.htm (Accessed 26 Mar 2017)

expect, quite reasonably, that anyone who is not an employee but has access for such purposes as deer management should hold personal third party liability insurance. At a more basic level, such as recreational stalking, the sporting insurance provided by a number of national organisations such as the British Deer Society, National Gamekeepers' Organisation or the British Association for Shooting and Conservation is quite sufficient. If, however, commercial activities (such as escorting paying stalking guests) is taking place, a more comprehensive commercial insurance policy may be required. Remember as well that if a paid assistant is employed, it may be necessary to ensure that they are covered by employer's liability insurance.

Do not neglect to ensure that suitable equipment cover is also in place. Even a few high seats can amount to some considerable capital value, before you even begin to take more expensive items such as vehicles and cold storage facilities into account. In addition, it makes sense to ensure that motor insurance covers your activities if you are venturing off-road regularly.

Grant aid

Grant aid in support of deer management activities is frequently available. Currently (in 2019) it is administered by the Forestry Commission and may provide help with activities ranging from the production of woodland management plans, woodland improvement, organised deer population management, or the creation of deer lawns and infrastructure to support culling. The sums involved can be significant; up to 80% of costs have been allocated in specific instances.[2]

Grants generally apply to woodland areas or Sites of Special Scientific Interest, or areas that are in close proximity to them. The actual grants available, and the form of their administration, may vary across different parts of the United Kingdom, and it is important to remain aware that once in the scheme landowners will usually be expected to carry out any recommendations arising from the assessments that take place before a grant is made.

As the grant aid packages are subject to change, it is advisable to contact the relevant local Forestry Commission office for advice and further information on eligibility and the implications of taking an application further.

Training

Unlike in many other countries, there is no legal requirement in the UK for those involved in the physical management of deer to hold any formal qualifications in the subject. However, as in all walks of life, their activities are covered by other laws, personal liabilities and a duty of care, and it follows that a degree of preparation for those activities is strongly advisable. Attending appropriate training courses is a suitable means of achieving this.

The cornerstone of training for deer stalkers in Britain is the Deer Stalking Certificate (DSC). Administered by a central national organisation, Deer Management Qualifications (DMQ) Limited, it is available at two levels through providers such as the British Deer Society, British Association for Shooting and Conservation and others. Whilst not a prerequisite for obtaining a Firearm Certificate or being granted

2 The Deer Initiative (2009), *Best Practice Guide – Grant Aid for Deer Management*; available from www.thedeerinitiative.co.uk/uploads/guides (Accessed 27 March 2018)

deer stalking rights, DSC1 and 2 are often regarded by police, landowners and others as suitable proofs of competence. They, too, have a duty of care and need to satisfy themselves that anyone authorised to access their land or possess and use firearms has provided such proof.

DSC1 exists to provide an entry level standard for stalkers. Typically delivered over three or four days of training culminating in final assessments, it is a mixture of theory and practice. While it is in no way arduous and can be taken by a complete novice, a small degree of preparation is advisable (most training organisations provide a comprehensive manual which can be read beforehand). The certificate involves five assessments:

▶ Written - a 50-question paper with multiple choice answers, covering deer biology and ecology, law, stalking techniques etc.
▶ Visual - a simple test of deer identification covering the six British species.
▶ Game meat hygiene - a 40-question paper, once again multiple choice, considering normal and abnormal deer behaviour, common diseases, signs of ill-health, hygienic working techniques etc. As with the written assessment, a pass mark of 80% is required.
▶ Safety - a combination of verbal questioning and practical demonstration of safe practices.
▶ Shooting - conducted on a rifle range to demonstrate a basic level of accuracy.

⬇ DSC1 is a combination of practical and classroom-based assessment.

These five assessments stand alone as steps towards the award of DSC1; once passed, they remain valid for three years from the date of registration and do not all have to be passed at the same time. When all five have been achieved, DSC1 will then be awarded. Like the DSC2, the DSC1, once achieved, is a lifelong qualification which carries no requirement for reassessment or refresher training to remain valid.

While some experienced deer managers may still question the need for them to undertake the DSC1, most will admit that they learn a great deal from it as well as attaining a qualification that is recognised both at home and abroad. In addition to this, the game meat hygiene aspect provides the successful candidate with the Trained Hunter status required to pass venison into the food chain.

DSC2 is the next stage of training. Portfolio based, it assesses the candidate's knowledge and practical competence when culling deer efficiently, legally and safely, as well as dealing with the resulting carcases hygienically. It is based on the stalking, shooting and subsequent processing of at least three deer whilst witnessed by a DMQ Approved Witness, who will pose any additional questions necessary. The resulting evidence is collected and submitted for evaluation before the award of the qualification. DSC1 is a prerequisite and, as it is a practical assessment, the candidate is strongly advised to obtain sufficient experience before attempting it. Once again the candidate has three years from the date of registration to collect the necessary evidence.

While DSC1 and 2 provide a sound basis for the training of a deer stalker, there is relatively little content addressed towards actual management as opposed to stalking. **Deer Managers' Courses**, such as those run by the British Deer Society, provide such knowledge and delve into the deeper issues. You do not need to be involved in actually culling deer to gain benefit from such a course, which will have value for anyone involved in the decision-making processes associated with deer and their issues. In addition to stalkers, they are aimed at landowners, land managers and anyone who may need to advise others, or those who simply want to expand their knowledge.

It is not necessary to complete the DSC1 if all you want to do is obtain Trained Hunter status. **Game Meat Hygiene** training is available from various sources, including various land-based colleges, the National Gamekeepers' Organisation and BASC. These tend to be just a single day in duration and have the additional benefit of qualifying the candidate to pass small as well as large game into the food chain.

If you, or those working with you, are to use any potentially dangerous machinery it makes sense to attend relevant training which leads to certified competence. For instance, while chainsaws can be bought over the counter and ATVs such as quad bikes need no licence to be driven off-road, formal training demonstrates a duty of care and may be required by insurers in any case.

All of the above can be classified as part-time courses. Very few people can afford greater time and expense unless they are in a position to undertake full-time education, and in any case deer management tends to be included as just one element of the longer courses (which are usually linked to wider game and wildlife management qualifications for conservation workers, gamekeepers and the like). Most of those with relevance to the subject that are available tend to be in further rather than higher education. The websites of the major land-based colleges will provide further information on what is available.

Should training be compulsory for deer stalkers and managers as it is in other countries? Whatever your views on this, it is important to recognise that wherever safety (whether it be from a misplaced bullet or contaminated venison) is concerned, some proof of competence is necessary to reassure not just the authorities but also the general public. Additionally, sufficient legislation exists to ensure a personal liability for ensuring that all tasks, whether they involve constructing a high seat or preparing a carcase for the larder, are performed to the highest standards. A professional approach is underpinned by knowledge, and by improving the latter through exposure to best practice our approach to managing deer can only be enhanced.

Covering costs

When you take the time, equipment needed and other aspects of managing deer into account, the costs can add up. Deer, however, are a very real asset which can pay their way if carefully considered. At the time of writing the venison market is buoyant; as recently as the 1990s it was estimated that some 70% of wild venison sourced within the UK ended up being exported to the continent, primarily France and Germany, but today many game-dealers complain that they are unable to obtain sufficient supplies to satisfy their local demand. Of the major supermarkets, Marks & Spencer sold three times as much venison in 2011 as it did in 2010, and by 2014 Waitrose reported that its venison sales were up 92% year-on-year. This trend shows no signs of slowing.[3]

The reason is not difficult to see: public recognition of the fact that venison is a healthy, ethical and sustainable source of meat has grown significantly, thanks to enlightened marketing and the publicity given to it by the rise of the 'celebrity' chef.

3 The Scottish Venison Partnership, *The UK Venison Market*; available from www.scottish-venison.info (Accessed 10 April 2018)

Venison is low in cholesterol and contains less than 2% fat (less than skinned chicken), and less saturated fats than other red meats. It is also a good source of iron, anti-oxidants, potassium, riboflavin, niacin and is high in vitamins B6 and B12. It is also higher in beneficial polyunsaturated fats than other red meats, with a highly favourable saturated to polyunsaturated (P:S) ratio and omega n-3 to omega n-6 ratios. This is largely due to the fact that deer feed primarily on natural vegetation rather than high energy cereals.

As far as the manager is concerned, this public demand for venison translates into a ready market with the game dealer. With prices that might range from around £1.50 to £2.50 per kilo, depending on species, condition and presentation of the carcase, stalkers can expect a realistic return for their efforts.

← *A group of gamekeeping students visit a game-handling establishment. Attention to hygiene is essential when passing any game into the food chain.*

→ *Letting stalking rights or hosting paying guests may bring benefits, but there can be drawbacks too.*

Revenue can, of course, be radically increased by selling prepared venison direct to the public. With venison loin currently retailing in some places at around £42 per kilo, roasting joints £24 per kilo and even the cheaper stewing cuts or mince at over £10 per kilo, there is potential for an even greater return but you do need to take the increased labour involved into account. There will also inevitably be increased infrastructure costs as premises will need to meet appropriate standards and you will need to register as a food business with your local authority. If you find this approach attractive, comprehensive guidance is available from the Food Standards Agency.[4]

While the venison produced has a tangible value, so too can the actual process of producing it have one. It is often tempting for a landowner to consider selling the stalking rights for their ground, but there are significant caveats attached to this. While the sums involved can be attractive (over £10 per acre in annual rent has been reported for some prime roe-stalking ground in the south of England), great care needs to be taken to determine the credentials of the person taking the lease. All too often one hears of unscrupulous tenants shooting out the resident deer, or allowing their sporting clients to shoot them, with little consideration other than simply obtaining trophies or venison income, and the result can be a significantly unbalanced population in a very short time. Needless to say, the matter of breeding females is seldom properly addressed. Equally dangerous is the sporting tenant who monopolises a large acreage with little hope of achieving the actual cull required to keep deer numbers under control. Poor management, for whatever reason, almost inevitably involves overshooting male animals, whilst the females are left to proliferate.

It is often far better for a landowner wishing to sell the stalking rights to set a modest sporting rent, or some other arrangement involving perhaps a share of carcase proceeds, in return for a tangible agreement regarding objectives while retaining a degree of control over how culls are conducted. Mutually beneficial working agreements on smaller management areas are often based on a 'one for you, one for me' share of the carcases and in many successful examples money is just a minor consideration. It is preferable to have a professional, humane and intelligent approach to deer management in the long run rather than a short-term financial gain which has the potential to incur far more problems than it actually solves. A sensible solution can be for landowner and sporting tenant to meet halfway, with a contract specifying a need for a cull plan which is to be formally agreed and then achieved each year.

Likewise, it is important to bear in mind that an obsession with trophies can be harmful to balanced deer management. In many areas with burgeoning deer populations, there are often very few mature male animals to be seen among the large numbers of females and immature animals of both sexes, thanks to the overshooting of the higher quality males. It has often been observed that many deer stalkers, faced with a mixed herd of deer all of which are in season, will almost invariably select the one with the best antlers. Restraint is essential and the occasional trophy should be viewed as a reward for the more essential business of culling breeding females, immature and old animals. Whilst it is true that antlered males make up part of a properly constructed cull plan, and older animals tend to have bigger antlers, it is also true that as a percentage of the overall cull they will form a relatively small element.

4 Food Standards Agency (2015), *The Wild Game Guide*, available from www.food.gov.uk (Accessed 11 April 2018)

Reporting

An annual report to the landowner is not just a sign that you are taking matters seriously; it also builds up a valuable historical record of progress and trends as the years pass. For this reason, although a verbal report has more value than none at all, a written one is preferable. It does not need to be a large document and can be tailored to suit local circumstances and the extent of deer management operations.

The timing of any report most usefully coincides with a recently conducted census so that the cull plan for the coming year can be included at the same time. It can also contain other important information such as the certified inspection of high seats and any review of risk assessments. In this way the production of any necessary paperwork becomes one simple annual exercise.

Many details in the report will not change and it will be a matter of simply updating material where necessary. The format need not be complicated, and it can be as sparse or involved as required. Comment on the previous year's achievements, deer numbers and health, any planned cull for the coming year and other relevant detail can all be included, although brevity is the key – excessively long documents tend to go unread, or at best are merely scanned, by those who receive them. A summary of what has been culled during the preceding year, easily maintained on a spreadsheet, should be included.

If required, the report might also be copied to other interested parties such as other land-users or members of a deer management group, but do be aware of the need to safeguard sensitive information and the requirements of the Data Protection Act. For this reason, and other sensitivities which may be attached to deer management activities, your report should be treated as a confidential document and its distribution carefully controlled.

An example of a simple report that should meet most purposes is at Appendix V. ■

5 High Seats and Other Infrastructure

Deer management on a small, localised level demands little in the way of infrastructure, but as efforts intensify so too will the need increase for tools to assist with the operation. This chapter will focus predominantly on the siting and design of high seats, an invaluable tool for effective management, but it will also draw attention to other matters that need to be considered.

Seats, high and low

Deer are supremely good at spotting movement, so it pays to be static and let them come to you. Some form of seat helps here, and of course raising it off the ground can also open up lines of sight where cover is thick. High seats, being several feet up in the air, are especially useful as they keep human scent well away from the woodland floor, and of course allow a stalker a safer backstop for the bullet where ground is flat. Whether for observation, photography or shooting, seats should be considered an essential element of infrastructure.

It pays to take care when selecting a new seat location: if you get it wrong, it is easy to waste a great deal of time, effort and money putting up a permanent construction. The first important question is – do the deer actually use the area? If you know your ground well you should have the answer but be aware that movement patterns can change from time to time.

You also need to consider longer term land usage before putting up a permanent seat. The chosen site may be good for now, but are any changes to the woodland or agricultural use going to move the deer elsewhere or put a seat out of action in some other way? Beware especially of young plantations; many elaborate box seats, erected to protect new plantings, have been rendered useless within a short time as the growing trees gradually eliminate the views of the ground in all directions.

Portable seats are a very useful solution in the short-term and also as a way of proving a site worthy of a full-time structure. They are quick and simple to erect, but take care in areas where there is public access. Being easily carried they are just as easy to steal, so make sure that they are secured to their supporting tree by a stout chain and padlock to deter the opportunistic thief. These days, sadly, cordless cutting equipment can make short work of even the strongest chain, so it is tempting providence to leave one in the same position for too long.

← *A poorly sited permanent high seat surrounded by conifers, which have grown rapidly and obscured its surroundings - a testament to short-sighted planning.*

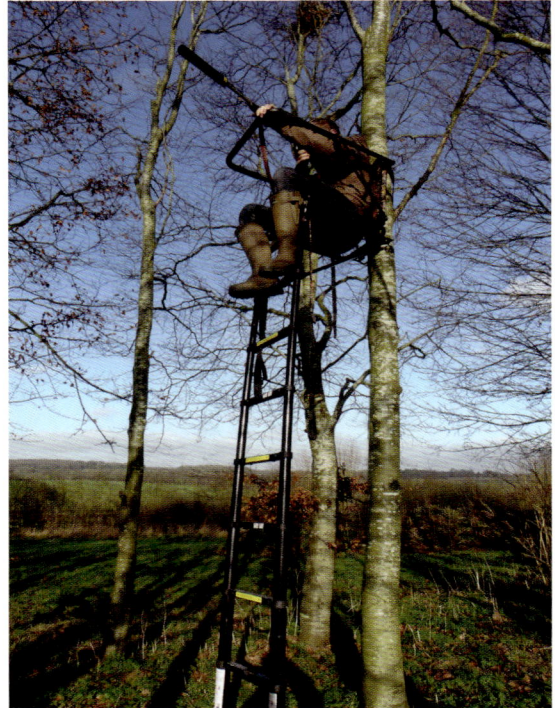

↑ ↗ *A portable high seat can be easily transported and is ideal for dealing with short-term problems or 'proving' a location for a more permanent construction.*

Once you have identified a clearing or small field that attracts regular visits by deer, it is time to start thinking about where the best place for the seat is going to be. If it is a lean-to type, you will need a suitable tree against which you can secure it. This needs to be large enough to prevent it from swaying in the wind, and sufficiently clear of branches to allow visibility. You also want sufficient background to break up or disguise your silhouette.

Whilst good fields of fire are important, do not look for big, uninterrupted vistas for hundreds of yards in every direction. Such places will discourage deer, as they may feel too exposed and thus unlikely to hang around in them for long. Small glades or scallops in a wood-line are much more suitable than wide open spaces. It is far better to choose a view with plenty of scattered cover, which not only offers browsing opportunities to attract the deer out but also allows them to relax. It does not matter if a deer is temporarily out of your view as it will move into a position offering you a shot eventually. By the same token, you do not need to see for hundreds of yards; restrict

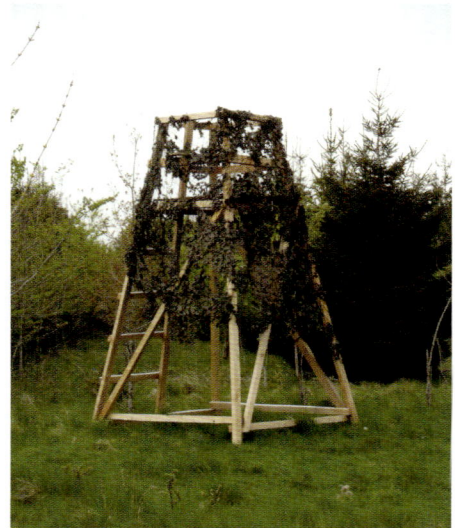

→ *A free-standing high seat placed to cover track junctions in a young plantation, where no mature trees are available for support. The camouflage net helps to screen the occupant from view.*

yourself to places where whatever you see will be within a realistic range for a shot. In practical terms this needs to be no more than about 150 yards, depending on a realistic appraisal of the competence of those you expect to be doing the shooting. There is no point in offering the temptation of long, risky shots. Make sure as well that there are safe backstops for the bullet if a chance does present itself.

Try to identify regular movement lines which are worth overlooking. You will often find well-used deer paths, or racks, which give away the habitual routes that deer take between their feeding and bedding areas. Hedge lines or other sources of cover in more open areas are well worth paying attention to as well.

If you are going to be overlooking a ride or a track, try to site the seat at a junction where you have views in more than one direction. If the undergrowth is thick between the trees, tracks will probably provide the only opportunity for you to see the deer as they move about within a wood. Wider rides with vegetation on each side may persuade the animals to linger rather than cross without stopping. If your ground holds a pheasant shoot, the deer will learn to visit the feeders, especially when natural forage is sparse during the winter, so having one or two of these within range of a seat can pay dividends. In the same way, game crops can also be productive.

When putting up the seat, angle it in such a way that the firing position will be as comfortable as possible in every direction. Right-handed shooters may find shots to their extreme right awkward and vice versa. Try instead to set the seat at an angle to the main shooting direction that suits you. If both left- and right-handed shooters will be using it, straight ahead is the best compromise. Do take care to avoid places where people may approach unseen, such as on footpaths or other public areas.

You must also identify a suitable approach route to the seat, preferably into the prevailing wind direction for your ground. If you fill the ground in front of the seat with your scent, you will give the game away before you even occupy it. An alternative route in for when the wind direction changes is also helpful. Keep your approach clear of new growth so that you can move into position silently, and don't cut the path in a straight line; put a dog-leg in it so that you are not in full view of the area you plan to overlook as you walk in.

Some kind of range-marking is especially useful in more enclosed areas where shooting chances may be fleeting. You may not have time to check with a rangefinder, and in poor light it is quite easy to mistake the distance to your target. Some people set out white-topped range posts at 50-metre intervals, or even paint the actual range onto the sides of trees, but different coloured splashes of paint on tree trunks and other available objects are just as effective – as long as the observer knows the colour code of course. Alternatively, you can fix range

← A continental design which is something of a hybrid between a high and a low seat, built on a triangular base. The upright planks form a removable seat and the overall structure is easily moved.

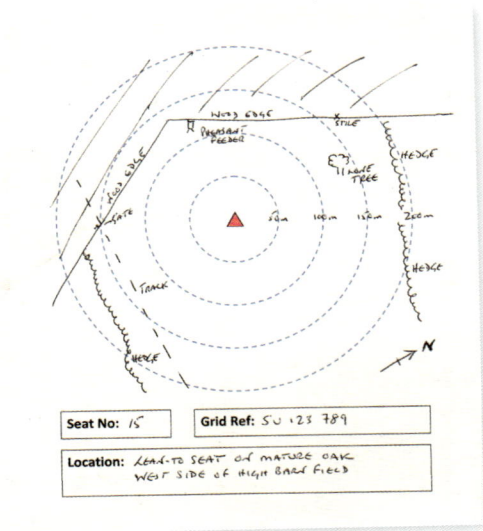

Seat No: 15 Grid Ref: SU 123 789

Location: Lean-to seat of mature oak West side of High Barn Field

↑ A simple range card, pre-printed and filled in by hand. The red triangle represents the seat location, and the circles can be marked to represent different ranges as required. Including a physical description of the location and a grid reference can be helpful to visitors who may be unfamiliar with the area or need to describe their position in an emergency.

↓ All high seats should carry a prominent sign warning against unauthorised access.

cards onto the rails of your seats - just laminate a few blank copies and complete them in permanent marker pen.

Finally, do not neglect the carcase extraction route. It is one thing physically packing out a roe or muntjac, but a bigger deer can be very awkward if you cannot get to it with your Land Rover or quad bike.

Seats and the law

A high seat is only a temporary structure and planning permission is not needed to erect one, especially if it is for the protection of forestry or agricultural crops, although of course you should consult the landowner first. You do need to be aware of a few obligations, however. If anyone, whether they are authorised or not, injures themselves when using one of your seats you may find yourself held liable, so it is necessary to pay due regard to its construction, siting and maintenance and to comply fully with health and safety regulations and other relevant legislation.

You will be able to reduce the risk of liability by taking a few sensible precautions. Ensure that the seat is properly designed and constructed. Treated timber rather than green material cut locally is best if you are building your own. It goes without saying that all materials which have to bear weight must be of a suitable strength; pay special attention to uprights, rungs, spars and flooring materials. Build in safety features such as wire-mesh stapled across wooden floors or ladder rungs to reduce the chances of slipping, and make sure that the rungs themselves are wired or otherwise reinforced against breakage. The seat itself should be stable and secure; if of a lean-to design, use rope or a ratchet strap, not nails or screws, to secure it to the tree.

Keep a written record of where all your seats are and ensure that you conduct regular inspections to confirm that they are serviceable. Wooden seats may rot over time, while metal seats can rust; welded joints need special attention, and paint may hide corroded areas. A formal annual tour of inspection is recommended in addition to more informal checks every time a seat is used. Any seat that is not fit for use should be repaired or removed without delay. A record of inspections and maintenance should be kept.

A removable ladder is a good way to prevent unauthorised use; another is to cover the rungs so that no part of the ladder can be accessed when the seat is not in use. It makes good sense to site a seat well out of public view, and away from footpaths or any other areas of public access. Finally, it is essential to ensure that every seat is clearly labelled with a suitable sign indicating that it is not for public use. Keep a spare supply of these in case they need replacing.

Building a lean-to high seat: a step-by-step guide

This is a simple but functional design, and the finished seat is strong, solid and takes two occupants comfortably. If you want only a single-seater you might reduce the front dimensions to 3 feet, but keeping it at full width makes it more stable and extends your options.

Pressure-treated wood is best, as it will help to extend the working life of the seat. Rough sawn wood is also fine, and certainly cheaper, but some form of preservative will be necessary and you will need to remember to reapply it regularly. By working in multiples of 4 feet you reduce wasted wood to a minimum, and the overall build cost comes in at around £50 – not bad for a seat that will give you many years of comfortable service if carefully maintained. Building it is at least a two-man job; an extra pair of hands is essential for some tasks.

Tools:
◗ Drill and appropriate bits
◗ Saw, hammer and fencing pliers
◗ Tape measure
◗ Bow saw and loppers (for clearing branches and undergrowth as necessary)
◗ Personal protection – hard hats, working gloves and protective boots

Materials:
◗ 2 x 12ft, 4in x 2in planks (4in x 3in is better but heavier) for front posts
◗ 5 x 12ft, 3in x 2in planks (to make 2 x 8ft and 11 x 4ft lengths)
◗ 1 x 5ft builder's plank (for seat)
◗ About 60 x 3½in/90mm screws (or an appropriate size for the thickness of the planks used). Screws are more secure, and much easier to reposition than nails if necessary.
◗ 2 x 6in coach bolts (optional)
◗ Low tensile wire and fencing staples
◗ Ratchet strap or rope
◗ If you are not using pressure-treated timber, creosote substitute or similar and a means of applying it

← Step 1 – Create the sides
Fix the 4ft side rail to the top of the front 12ft post and the supporting diagonal. Make sure that the top rail and the front post are at 90 degrees to each other, then secure the bottom of the diagonal to the front post. Now fit the extended seat support rail in place about 22in below the top rail.

Continues on next page ➡

← *The opposite side should be a mirror image of the one that you have just made. Don't skimp on screws anywhere on the seat; use at least two on each joint to make it secure. For increased strength, the base of the diagonal piece can be fixed to the front post using coach bolts.*

→ **Step 2 – Connect the sides**
Connect the two side pieces with 4ft top rails, and a back rail just below the seat support (the bottom rail shown is only there to square off the sides, and will be removed later). The rungs will provide the remainder of the front bracing.

↓ **Step 3 – Attach the rungs**
The rungs are now fixed to the front posts. They need to be fixed to the <u>inside</u> of the posts as that is where you'll be climbing up. The positioning of the topmost one is crucial as it will act as a footrest while you are sitting, so it should be at a comfortable height. About 14in or so below the level of the seat is about right. Space the remaining rungs evenly some 16in apart – don't be tempted to place them too far apart as it makes climbing awkward.

↑ **Step 4 – Tree supports**
Two diagonal struts at the back of the seat support are a useful addition; the angle of these can be adjusted to fit the width of the tree supporting the high seat.

➡ **Step 5 – Wire the rungs**

Using the fencing pliers to tension the wire, staple it to the ladder rungs and front posts in a continuous zig zag pattern. Two sets of hands are useful here: one person tensions the wire whilst the other fixes the staples. This is an essential safety precaution against a rung breaking under the weight of a climber.

⬅ **Step 6 – Position the seat**

Locate the seat against the chosen support tree and secure it with rope or, better yet, a ratchet strap. For safety, make sure that you have a helper or two holding the whole thing steady as you do so.

Add the builder's plank as a seat (an offcut of wood at either end prevents it from slipping off its support rails), and affix the warning sign to deter unauthorised users.

➡ *The finished seat in position, overlooking an area of new plantation. A scrap of camouflage net or similar stapled around the sitting area would be a useful addition to help mask the occupants from view. Note that the ladder section is set vertical rather than leaning against the tree. Entry is by climbing up the inside.*

Building a free-standing high seat

Sometimes there is no convenient tree against which to support a lean-to seat, so the structure needs to be free standing. The following design is based on a continental shooting stand and can be easily adapted to meet local circumstances using the available materials. It is very solid and does not need any bracing wires as the legs are splayed to ensure stability. The key to construction is not to skimp on any weight-bearing areas. If a roof is desired, which will usefully provide both shelter from the elements and shade to help conceal the occupant, this can simply be added afterwards, or the structure could be based on longer upright posts to make the roof an integral part. Camouflage netting can replace the more solid side-cladding on the example shown if desired. It is best to build such a seat on site because its overall weight, whilst enhancing its stability, can make transport an issue.

Tools:

▶ Hammer, wood saw, bow saw, ratchet set (if using Timberlock fixings), cordless drill.
▶ Personal protective equipment (hard hats, gloves and working boots).
▶ A ladder is useful for any finishing work necessary once the seat is upright.
▶ **At least one extra pair of hands – this is definitely not a one-person job.**

Materials:

▶ 4 x 12ft, 4in-diameter full round fencing posts
▶ 12 x 12ft half round fencing posts
▶ Flooring (builders' planks, or solid sheets on supports of 3in x 2in timber: both work well)
▶ Timberlock screw bolts or similar fixings (e.g. threaded rod and nuts or coach bolts)
▶ Fencing wire and staples (for ladder)
▶ Wire mesh (weldmesh or similar) for the platform floor
▶ Supports and plank for seat (around 6in to 8in wide for comfort)
▶ Cladding for the side – 36 thin planks stripped from 6ft pallets and cut to size (chipboard, tarpaulin, camouflage net or similar are good alternatives)
▶ Nails or staples for fixing the cladding
▶ Creosote substitute and brush (for any untreated wood)

⬆ Step 1 – Construct the front and back sections
The front and back of the seat are constructed as two identical A-frames while still flat on the ground. The top rail should be about 4ft long and the base rail needs to be around 5ft 6in to ensure a stable base for the finished seat. The platform rail is positioned about 40in below the top rail. The platform rail at the base of the seating area should be on the inside of the frame to provide a support for the floor. For all main fixings, Timberlock screws or similar are recommended as they can be easily removed if readjustments are necessary. Otherwise coach bolts or threaded rod, cut to length, are better and improve strength.

⬆ Clad the front frame of the seat with pallet planks or similar and cut to fit if camouflage netting is not being used (this can be added once the seat is complete) as parts of the construction are more easily reached when it is on the ground.

← Fit a ladder to the rear frame, allowing about 16in between rungs. Wire these as described for the free-standing seat.

↓ At this stage, complete any application of wood preservative as the higher parts of the construction are more easily reached when it is on the ground.

← Step 2 – Connect the two completed A-frames

Join the front and back sections together with top, platform and base rails of the same lengths as those on the A-frames. Add bracing spars at the front and sides to steady and strengthen the seat. At this stage the construction can be set upright (a little forethought means that it is simply tipped backwards into its final position).

← Step 3 – Construct the platform floor

3 x 2in fencing rails, laid front-to-rear across the platform, provide a very strong support for 5-ply boards cut to size. Scaffolding planks would be just as suitable. It is important that, whatever you use, they are suitable for serious load-bearing and inspected for safety regularly. Wire mesh stapled to the surface is a sensible precaution as wet wood can be slippery underfoot.

Continues on next page →

← **Step 4 – Complete the seating area**
Once the floor is in place the sides can be clad.
For seating, fix rails to the sides at an
appropriate height for a piece of scaffolding
plank to be laid between them to form a seat.
A simple but effective innovation is to fit a 1in
scrap of timber to each end of the seat plank,
so that it can be turned over and the
additional height will then accommodate a
slightly longer-legged occupant more
comfortably.

→ *The finished high
seat. As for all
wooden seats, a
small piece of
breeze block or
paving slab placed
between the
bottom of each leg
and the ground
allows water to
drain away from
the wood and this
will prolong its
working life.*

Building a low seat (doe box)

Most stalkers who regularly cover the same ground have their favourite 'sitty places', usually on a piece of raised ground which gives them a good field of view and safe backstops for shooting opportunities. As movement is probably the biggest giveaway of your presence as far as deer are concerned, managing to keep still and hidden immediately gives you an advantage. Having a box to sit in doesn't only conceal you from sight – it is infinitely more comfortable than sitting or lying on a damp grassy bank for a couple of hours or more

The low seat, sometimes called a doe box or just a shooting box, is quite simply a permanent ground-level hide for the stalker. As it is not built several feet off the ground, there are no major load-bearing issues, safety concerns are greatly reduced and of course you do not have the problem of finding a suitable tree to place it

← *A doe box or low seat, constructed almost entirely from materials salvaged locally.*

against. In fact, as the stalker is contained within the structure, you have few of the problems associated with concealing yourself in a high seat. Contrary to popular opinion, deer *do* look upwards, as anyone who fidgets in an open high seat will quickly discover! A box, on the other hand, has many advantages and even allows those who stalk with larger breeds of dogs to accommodate them as well, rather than having to leave them in the car or at the base of a seat.

With the exception of four fencing poles, which form the corner supports, and an old plank for the seat (an old scaffolding plank is ideal), all of the wood required for the example pictured was salvaged from a stack of unwanted pallets, making it a very low-cost project indeed. Proofing the structure with creosote substitute or similar ensures that it blends in properly with its surroundings and has a long, useful life.

Tools:

◗ *Hammer, wood saw, bow saw, pile driver, spirit level, screwdriver, brush for proofing*

Materials:

◗ *4 x 7ft fencing posts*
◗ *About 45 x 150cm pallet planks (front width and roofing)*
◗ *About 35 x 100cm pallet planks (side width)*
◗ *Pallet cross pieces (foot rail, seat support, roof rails and door frame, cut to size as needed)*
◗ *Plank for seat (around 6in to 8in wide for comfort)*
◗ *50mm (1in) and 75mm (1½in) nails*
◗ *Hinges, hasp and staple, padlock and fixing screws*
◗ *Creosote substitute*

Construction is straightforward. It is essential to ensure that the first of the four corner posts is driven in truly vertical, otherwise the entire structure will be off-centre; then subsequent spacing is determined by the length of the cladding planks available. The two front posts need to be driven in about 2 inches more than the back ones, to ensure a sufficient angle for rain to run off the roof.

The simple bench seat should be about 19 inches off the ground; this should be comfortable for occupants of all heights. It should not stretch across the width of the box but be supported by a horizontal beam at the side opposite to the entry, finishing on a short post about a foot from the door to allow easier access and for the stalker to slide around to shoot out of a back port (if there is one).

Trial and error will determine an appropriate height for the viewing ports and supporting rail. Keep the viewing/shooting ports as narrow as possible, just sufficient to enable good visibility and enough space to slide a rifle or camera out quietly. Too wide a shooting port, and you are unnecessarily exposing movement to the quarry. If ports are needed at both the front and back of the box, you run the risk of the occupant being silhouetted against them. A couple of pieces of camouflage netting will help to prevent this, as long as the mesh is wide enough to slip a rifle muzzle through easily. You might also consider fitting a foot rail to the box interior – being able to keep your feet off the ground on frozen mornings may mean all the difference between the discomfort of frozen toes and keeping warm.

To make the box lockable against casual passers-by, a basic door can be added with a hasp and staple for a padlock (a sheet of old tarpaulin that can be pulled aside and then hooked back into place is also effective but not so secure). A roof is an important part of the overall structure not, as you might suppose, for keeping the occupant warm and dry, but to create shadow within the box to conceal the occupant. Of course, being warm and dry is always a big bonus. Pallet cross-pieces across the insides of the front and back posts provide a solid support on which to fix the roofing planks. Roofing felt or tarpaulin prevents annoying internal drips if it is raining.

The great advantage of this type of seat is that there is no need for a formal plan and you can build to the size of the materials available, making it up carefully as you go along. This means that you can use whatever materials are to hand and keep costs down to a bare minimum.

The design pictured will easily accommodate two people, though you can adapt dimensions to the length of wood cladding you have. You might even

1 Home Office (2013), *Surveillance Camera Code of Practice*, London, The Stationery Office; available from www.assets.publishing. service.gov.uk/government/ uploads/system/uploads/ attachment_data/file/204775/ Surveillance_Camera_Code_of_ Practice_WEB.pdf (Accessed 7 August 2018)

consider having slightly lower sides and a folding canvas roof rather than a solid wooden one if the box is in a good pigeon-shooting location, or next to a flight pond, or put it onto a solid base for moving between locations with lifting arms rather like a sedan chair. The only limit is your imagination.

Trail cameras

A relatively new addition to the deer observer or manager's armoury, the trail camera provides eyes on the ground 24 hours a day, long after you and others have gone home, taking pictures conventionally or using infra-red technology. With digital video and sound available in addition to still photographs, such cameras are truly versatile and offer date and time stamping options to show the user exactly when the camera was triggered. More advanced models can even transmit pictures to the user's mobile phone or computer in real time.

While it is possible to spend a great deal of money on trail cameras, relatively cheap models are readily available which still offer reliability and excellent picture quality. Site them carefully, remembering that it takes a short time for the camera to wake up and trigger, so they should be set at an oblique angle to a trail rather than at right-angles to it – otherwise you will end up with pictures of fresh air as the animal has already passed by. However, intelligent use and regular checking allows you to build up a good picture of what inhabits your ground, and you may come to recognise individual animals and their movement patterns. Working times are often longer than you might expect with excellent battery life, and solar-powered options are also available. Most operate with memory cards which can be swapped with spare ones whenever you happen to pass by.

Sometimes you will discover that the deer simply do not use an area regularly, perhaps saving you the wasted labour of erecting a high seat, or fruitless hours of sitting in one. They will also identify intruders, human and otherwise, for which reason gamekeepers and other country workers find them invaluable in places where security may be an issue.

A word of caution is in order. Even the cheaper models are still valuable, and to deter theft they are best placed well away from public access areas and, where possible, secured to the tree they are mounted on by a strong cable and lock. I have found that a small laminated card, mounted just above the camera, stating *'This is a wildlife observation camera, please do not move it - your picture has already been transmitted to the base station'*, and including a contact number, helps. You may well find that you take a number of pictures of people leaning in close to read the notice but subsequently deciding to leave the camera alone.

There are few legal implications where trail cameras are used on private land, but there are special considerations when they are deployed in public places. A code of practice, which can be accessed on the internet, is available from the Home Office.[1] Be aware, too, that cameras are sometimes used covertly by people attempting to obtain evidence of illegal or other activities without the landowner's knowledge.

↑ An inexpensive but effective camera purchased as a 'weekly special' from a major supermarket.

↑ Trail cameras allow you to record the comings and goings of a wood's inhabitants by both day...

... and night.

Night vision

While night vision devices have no place in the shooting of deer, they can be of great assistance when conducting censuses, for general observation purposes or for locating shot deer once darkness has fallen. The latter consideration is important as, due to their crepuscular habits, most deer tend to be shot close to first or last light. Spotlighting used to be the only method available for night vision, using beams of white light in conjunction with binoculars to render deer visible. While the light signature can be reduced by the use of coloured filters, the beam remains highly conspicuous and can potentially disturb deer to the point that they become reluctant to leave cover. It is also, of course, more obvious to anyone else in the vicinity and may lead to public concerns.

There have been considerable advances in other technologies comparatively recently, and the old days of 'starlight scopes' – which relied solely on ambient light to provide a fuzzy monochrome image that often struggled to identify a man-sized object at a range of 30 metres or so – are long past. Once only really available to the military and other official users, it is now possible to purchase simply operated and effective handheld devices on the open market at relatively low prices. They are usually offered in monocular or binocular format, or as goggles which can be attached to the head with a harness strap. Effective ranges will vary between models, as will the option of being able to adjust the lenses for zoom or clarity, and this will inevitably be reflected in the overall cost. Whereas many older devices had specialist power sources that were not readily available on the open market, modern equivalents tend to be rechargeable or use standard batteries.

Modern image intensification devices no longer rely on ambient starlight or moonlight alone, but usually work in conjunction with projected infrared light, which increases their range and clarifies the image; many are offered with integral illuminators. Their effectiveness still tends not to be particularly good beyond limited ranges, though, especially where there is ground cover.

Photograph: Yukon Advanced Optics Worldwide

➔ A hand-held thermal imager capable of identifying deer up to some 800 metres away in absolute darkness.

Thermal imaging (TI), on the other hand, works by picking up heat signatures and has no reliance on ambient light, so it is effective in pitch darkness as well as fog or rain or even daylight. Any warm object, such as a living animal, shows up as a bright image against a darker background and, with practice, the operator can even learn to differentiate between the deer species and determine their sex. Even a relatively affordable model should offer an effective detection range of around 800 to 900 metres, while those at the higher end of the market claim effectiveness to several kilometres.

Photograph: Yukon Advanced Optics Worldwide

A fallow buck in velvet, viewed through the thermal imager shown in the picture on the page opposite. The tree trunk and fence are reflecting residual daytime heat.

Unlike some methods which rely upon projected light, TI is entirely unobtrusive. Where there is cover, such as in woodland, the deer may not always be entirely visible although it is surprising just how effective the devices can be when looking into even relatively dense undergrowth. A shot animal lying in cover will retain its body heat for a long time and can be found with relative ease, whereas conventional white light sources often fail to pick one up if it is camouflaged by branches and low growth. Using TI it is even possible to follow fresh blood trails.

Modern TI equipment is lightweight compared to its predecessors and has now been reduced to sizes that will fit in a coat pocket. Most models offer a long battery life and are designed for robust field use (although care still needs to be taken to avoid excessive knocks and scrapes). These and the rather less costly image intensifiers still represent a large outlay, however, and are not for everyone. It is always advisable to seek specialist advice before deciding to invest.

Tracking dogs

Anyone involved in the shooting of deer should have access to a dog that is capable of following up a wounded animal that needs to be located and despatched. Hopefully such situations will be rare, but even the most experienced stalker will admit that occasionally things may not go quite as planned. Occasionally the quarry, even if mortally hit, may not collapse on the spot but travel a short distance to expire quickly in cover close by, and even a large animal lying in relatively sparse cover can be surprisingly difficult to find for a human relying only on the naked eye. Otherwise a deer carrying even a relatively serious but non-lethal injury is capable of travelling very long distances while leaving little blood sign, and only a dog trained for the purpose will enable the stalker to find and deal with it. Not to make every possible effort to recover any animal, mortally hit or otherwise, should be unthinkable to any conscientious deer stalker.

In theory, most breeds of dog can be trained to be capable of conducting simple follow-ups. 'Hot' trails, made recently, are easier to follow and many deer stalkers own multi-purpose dogs which perform well in the shooting field as well as deer-stalking companions. Training such dogs is relatively simple, and most catch on quickly to the connection between finding a dead or moribund deer and a reward. A steady dog can accompany the stalker and be of great use in indicating unseen animals in cover long before human senses can locate them. Otherwise it is useful just to have a dog readily available if it suddenly becomes needed.

➜ *All deer stalkers should have access to a dog capable of finding a lost animal. Here a springer spaniel locates a roebuck which has run on a short distance before falling.*

↓ *A dedicated deer-tracking dog, in this case a German wirehaired pointer, being prepared for work. Such a dog will be able to follow a long-cold trail, often for many miles, before restraining an injured deer if necessary for despatch.*

Photograph: Tony Lowry

For more complicated tracking tasks, especially those 'cold' trails which have been left for a longer period of time or involve covering greater distances, a tracking dog trained solely for the purpose may be necessary. Such a dog will have more specialised skills. It will need strength and stamina, be able to indicate a find to its handler, and even be capable of restraining a mobile deer until the handler can effect the despatch. Like any working dog, it will need regular practice and revision training to keep its skills at the highest level. On the continent there are regular field trials, and dogs may be required to demonstrate their abilities of increasingly demanding artificial trails before they can be registered as qualified tracking animals.

Interest in the training of tracking dogs has grown recently in the UK, and a number of voluntary organisations now offer call-out services across the country. Most base their methods on continental practices, as there is a longer tradition of training dogs for work with big game overseas, and indeed some countries require possession or access to tracking dogs by law for those hunting large game. These organisations usually offer their services for free, or charge only expenses, such as travel. If you do not have a dog of your own it is well worth making a preliminary contact with a local member of a tracking group and storing their contact number on your mobile telephone against the occasion when their services may be needed. Some contact details can be found at Appendix II.

Carcase extraction

While the extraction of smaller deer carcases does not present so much of a problem, moving larger deer from the field back to more hygienic facilities for storage and preparation for the food chain needs to be given careful consideration in advance of culling operations. A purpose-made rucksack with a removable liner for cleaning will easily accommodate a couple of muntjac or a large roe carcase and, depending on the fitness of the user, can be comfortably carried back to a vehicle. Larger deer are not so simple. While dragging can be an option, it may expose the carcase to unnecessary contamination from the ground and other sources, so it is important to protect it from these.

⬇ Sometimes there is no alternative to dragging a carcase, although care must be taken not to expose it to potential sources of contamination. A 'drag bag' or sled would help to protect it, and a harness would make the stalker's task easier.

Deer sleds and drag bags are readily available commercially, but for ease of use the single-handed deer stalker will find a 4 x 4 or all-terrain vehicle more convenient. Winches and cradles attached to these not only enable easy transportation but can actually assist with loading and preparation of the carcase for the larder; all carcases need to be eviscerated or 'gralloched' as soon as possible after killing to allow them to cool as rapidly as possible, which will ensure that the resulting venison is of the best possible quality. A winch fitted to a vehicle, or a simple hand-pulley system attached to the branch of a convenient tree, can be of great assistance when handling larger carcases.

All bags, sleds or vehicles involved in carcase extraction must be kept scrupulously clean to avoid cross-contamination. If a carcase has been carried in any kind of waterproof sack it should not be left in it for any longer than is necessary, as it will sweat and fail to cool rapidly. Plastic trays or other liners for load-carrying areas are ideal receptacles for carcases carried in vehicles, as they are easily cleaned.

In all cases, care needs to be taken to avoid injuries related to lifting or dragging, and mechanised assistance is always preferable. If it is not available, the single-handed stalker needs to consider how assistance can be summoned should it be required. A dead deer can be an ungainly object to handle, and care must always be taken to avoid injuries inflicted by antlers and hooves while it is properly secured for transport. Take care, too, not to overload any vehicle and create a danger of tipping; this can be a particular hazard when using ATVs.

Deer larders

The scale and size of any deer larder or storage facility will be very dependent on the volume of carcases that it is expected to handle and how long they are expected to be stored there. Where carcases are taken directly to a game dealer with only the occasional one retained for personal use, this is rather less of an issue. Even then, though, thought needs to be given to basic hygiene, and higher ambient temperatures and the presence of flies and other vermin can quickly render a carcase unsuitable for human consumption. Whatever facilities are used they should be easily cleaned, allow the carcases to hang freely without touching walls or floor, and most importantly allow them to cool to 7°C or lower (small game, stored with internal organs in place, needs to cool at least to 4°C).

For occasional personal use with small deer, a simple chiller cabinet may be suitable. Glass-fronted refrigerated cabinets, of the sort seen in shops and filling stations displaying cold drinks, can sometimes be obtained reasonably cheaply during refits and equipped with a hanging rail. Thermostatically controlled and easily cleaned, they are ideal for occasional use and a standard 6-foot unit will easily accommodate two roe.

For more regular use or greater volumes of carcases, purpose-made game fridges or larger units are commercially available, or existing space in an outbuilding can be converted for the purpose and fitted with a suitable refrigeration unit. Whatever you opt for, it is absolutely essential that all surfaces are easily cleaned and that there is suitable drainage; that premises are adequately lit; and that all materials used for storing and handling are made of impervious materials which can be easily sterilised.

➡ A converted drinks chiller can be an ideal solution for the storage of carcases for personal use.

Many facilities will be used simply for the basic preparation and storage of carcases 'in the skin' before they are passed on to an Approved Game-Handling Establishment (AGHE) or venison dealer, in which case the requirement for good hygiene is fairly straightforward. If the carcases are to be sold on to others, including game dealers, the larder will need to be registered with the local authority. Such use is fairly common, and where large volumes of carcases are collected arrangements are usually made for the

AGHE operator to collect them at regular intervals. If, however, carcases are to be skinned and butchered on the premises, AGHE status may be required; further guidance is available from the Food Standards Agency who are also responsible for approving them.[2] If the products are to be supplied direct to the final consumer, it will probably also be necessary to register as a Food Business Operator.

Proper attention must be paid to appropriate waste disposal. While incineration is the best option, this may not always be practical for smaller operators but, whatever the case, larder waste should not be included with domestic rubbish and special arrangements will need to be made (see also Chapter 10). Currently there is a legal requirement for such waste to be stained to prevent it illegally entering the food chain. As regulations may vary between local authorities, it is recommended that the local council environmental health department should be contacted for advice on waste disposal.

For all venison entering into the human food chain, it is a requirement that it is inspected by a Trained Hunter (see Chapter 4) and tagged before the carcase is placed into cold storage pending further processing. The declaration tag (which can be obtained from the BDS, NGO and other suppliers) bears a unique reference number and shows the date, time and location that the animal was shot, along with other details which include any abnormalities noted. It is signed by the Trained Hunter and will also bear their personal identifying reference.

A simple HACCP system is advisable: a Hazard Analysis Critical Control Points protocol applies across the food industry, and simply looks at stages in production where there is a danger of issues arising and puts procedures in place to minimise them. Where deer are concerned, the stages might include careful scrutiny of the living animal, inspection and preparation of the carcase, extraction, storage, processing and further storage of the final venison products. Once again the Food Standards Agency offers a simple guide.[3]

As the traceability of carcases is essential, larders should maintain records of all those that pass through them, showing as a minimum the date, the name and number of the Trained Hunter, and carcase details by species, sex and weight. Additional records that are advisable might also include cleaning, larder inspections for cleanliness and serviceability, and daily temperature checks. ■

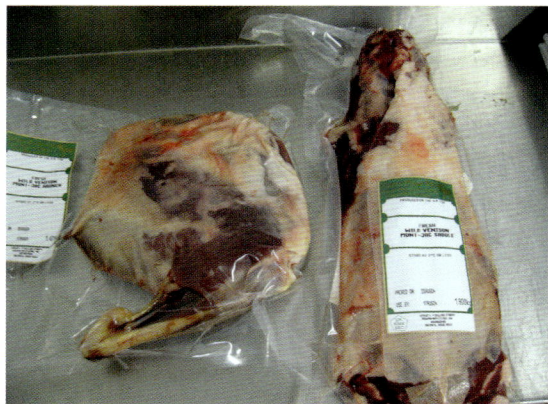

↑ (Top) Carcases held in a cold storage unit awaiting collection by an Approved Game-Handling Establishment.

(Bottom) Care must be taken throughout all steps of food production to ensure that the venison arrives at the point of supply in the best possible condition.

2 Food Standards Agency (2016), *Meat Industry Guide*; available from www.food.gov.uk/business-industry/meat/guidehygienemeat (Accessed 12 April 2018)

3 Food Standards Agency (2018), *HACCP*; available from www.food.gov.uk/business-industry/food-hygiene/haccp (Accessed 12 April 2018)

6 The Census

Why and when to census?

You cannot begin to formulate a strategy for managing deer, or any other wildlife for that matter, without knowing how many your ground actually holds. An effective census will not only provide you with numbers, but also the age and gender structure of the herd (and I use that term in the loosest sense for the less social species). Historically, comparison with past census results will give you an indication of just how effective the management strategy has been over the years.

Deer censuses have always tended to take place in March for a number of reasons. For roe, this used to be the month between the closing of the doe season and the opening of that for bucks, but the extension of the former in 2007 changed that – although in many cases it is more than possible to achieve the female element of the cull by the end of February. This not only removes the need for the distasteful task of shooting animals in a more advanced state of pregnancy, but allows a clear month for tasks beyond the population count, such as ride clearance, high seat maintenance and all the other things which can get sidelined as the buck season reopens in April.

For most species in southern Britain, March is the ideal census month for a number of more practical reasons. Firstly, the woodland understorey is still relatively open after the winter and vegetation has yet to start growing through again, making it much easier actually to see the deer. Secondly, and probably more importantly, you will obtain a more accurate impression of true deer numbers. Even in the less harsh conditions of the south of England, winter mortality can be high, especially among younger or weaker animals. One Dorset study noted that of 58 identified roe kids, 10 died during their first year of life, 5 simply disappeared and another 21 emigrated out of the study area.[1] Elsewhere it is suggested that of roe twins or triplets, very often only one will survive to the end of their first winter.[2] Further north, mortality might be even higher. Counts taken before or during mid-winter may therefore give a very false impression of future numbers; by March it should be possible to obtain a far more realistic picture.

In the north, it might make sense to delay the census until rather later. Here, of course, in more open areas visibility is not the major issue that it might be in woodlands obscured by spring growth. With more persistent inclement weather and poorer feeding, it has been found that winter deaths from malnutrition of red deer in Scotland might not occur until as late as April.[3] Male deer, having greater nutritional requirements, can often form a higher proportion of winter casualties.

1 Johnson, A.L. (1982), *Notes on the Behaviour of Roe Deer at Cheddington Dorset* 1970-80, Forestry Commission R&D

2 Whitehead, G.K. (1993), *The Whitehead Encyclopedia of Deer*, Shrewsbury, Swan Hill Press

3 Clutton-Brock, T.H. et al (1982), *Red Deer, Behaviour and Ecology of Two Sexes*, Edinburgh University Press

A need for caution

Before you start to conduct your census, a word of caution is in order as it is unrealistic to expect exact results. Deer are notoriously difficult to count, even under the most controlled of conditions. When I was teaching wildlife management students, I would often ask them to try to count the college herd of sika, which were contained in a largely open paddock only a few acres in size with the deer in plain view. Even then, estimates of overall numbers, let alone breaking them down by gender and age group, would inevitably vary wildly.

With free-ranging wild deer, matters become even more complicated. A census conducted at Kalø in Denmark is frequently cited as a classic example of under-estimation. There, as part of a scientific study, a total of 70 roe were assessed by experienced observers as being the total population in a set area of wood and farmland before it was deliberately shot out. By the time this process was complete, no less than 213 deer had been culled.[4] A similar exercise in Poland counted between 30 and 40 deer against an actual cull of 146.

Closer to home, Prior describes the results of a deer count on Wiltshire downland, a landscape dominated by grassland with relatively sparse cover compared with more dense woods elsewhere. A purely visual census initially suggested a population of no more than 35 roe. A subsequent, more comprehensive effort, using a combination of ground observers and a helicopter, actually suggested a springtime population of between 180 and 220 animals every year across a six-year period.[5]

On one moved census, involving a line of beaters gently pushing deer through a counting line, I recall driving into the mostly enclosed area where it was to take place in order to make final arrangements before everyone else turned up. The species on the ground were predominantly roe with a moderate population of muntjac, but in an open field surrounded by woodland I passed eight fallow, easily identifiable as a regular group by their sex and colour variety mix. By the end of the

➡ *Some deer can be more difficult to count than others. For such a large animal, fallow can be particularly elusive if the terrain suits them.*

4 Andersen, J. (1961), *Biology & Management of Roe Deer in Denmark*, La Terre et La Vie 1 41-53

5 Prior, R, (1995), *The Roe Deer – Conservation of a Native Species*, Shrewsbury, Swan Hill Press

day, no fallow had been recorded by the move process; an hour after the count had finished the small herd was again out and grazing not far from where I had seen them nine hours earlier. As an aside, although the numbers of roe recorded during the same day were much as expected, only three muntjac had been seen. Sometimes, despite the best of intentions, deer simply will not cooperate!

Different census methods lend themselves to differing circumstances: what works for the open hill with wide, uninterrupted vistas may not necessarily be appropriate for close woodland where the observer can see no more than a few metres at a time. Whilst modern technology may well make the process easier and more accurate, it is not always available or financially viable.

From the outset it is vital to recognise that census results may not always reflect the true picture. Be prepared to be flexible in your approach should the situation turn out to be different from what was first forecast. Remember: deer management cannot be treated as a precise process - it is a combination of science, reasoning, observation and common sense.

Identifying and ageing deer

There is little point in trying to conduct a census of any animal, bird or plant species if you don't know what it looks like! Very often you do not have the chance to examine a deer at leisure or at close quarters but, with familiarity, the ability to recognise a particular species from a fleeting glimpse or an impression of movement will develop. The experienced observer will not just say 'it's a deer' but rather 'it's a male roe deer'. For census purposes this skill needs to develop even further, to being able to identify whether the animal is young, mature or old, and possibly even what condition it is in - just in the few seconds available as the animal trots across a woodland ride.

In many respects, deer show signs of ageing in similar ways to humans, although, just like humans, some defy the rules. Antler development can be especially misleading and it frequently helps to disregard them entirely, considering other factors first. I distinctly recall watching what I thought was a yearling roebuck according to all the visible signs of antler development, build, attitude and behaviour. After it was subsequently culled I was staggered to find, on looking at its teeth, that it was an extremely old animal indeed, to the extent that it was a minor miracle that it had survived the winter. Such experiences are fairly common, but one hundred percent accuracy is not required for the purposes of the census, where a sensible overview of the population age structure is all that is needed.

Beyond obviously immature animals in their first year, for which birthing dates are well known, it is usually impossible to determine precisely the age of a living, grown animal (unless it has been visibly tagged at birth), so it is far better to think in more general terms of young, mature and old.

While there are many resources available that provide assistance with identification, the table on the next page offers some general guidelines on determining the age of a deer though it must be stressed that *not all signs are absolute*. Beware in particular that animals in their thicker winter coat may look misleadingly more heavily built than they actually are. It should also be noted that there may be some minor variations between species and, as the different species have different lifespans and effective breeding ages, age groups should be taken as:

	Large deer species	Roe and muntjac	Chinese water deer
Young	Not yet 3 yrs	Not yet 3 yrs	Not yet 1 yr
Mature	Not yet 9 yrs	Not yet 7 yrs	Not yet 5 yrs
Old	Over 9 yrs	Over 7 yrs	Over 5 yrs

	Young	Mature	Old
SEASONAL			
Coat change	Earliest	Later	Latest
Antler casting and growth	Latest	Earlier	Earliest
APPEARANCE			
Antler form	Light and simple	Fully developed	Reduced, thicker beam, 'going back'
Canine tusk (CWD bucks)	Not visible	Visible	Visible
Body	Slim	Muscled	Heavy
Neck	Thin	Firm and muscled	Heavy
Back	Straight	Level, slight dip	Dipping
Belly	Taut	Dropping slightly	Sagging
Rump	Narrow	Rounded	Well filled (but may become bony with advancing age
Facial expression	Trusting	Alert and wary	Suspicious
GENERAL			
Gait	Light and brisk	Deliberate	Stiff
Behaviour	Playful, curious and unwary	Purposeful	Cautious
Reaction to potential threat	Incautious, may investigate	Alert, staring and quickly reactive	As mature (very old animals may become less alert)
Feeding	Incautious	Alert	Very wary

Young   *Mature*   *Old*

↑ *Body forms showing the ageing process in roe deer.*

(From: *The Roe Deer* by Richard Prior, reproduced with the kind permission of the author)

The rolling census

There is nothing formal or complicated about a rolling census: the term simply refers to the day-to-day observation of deer on the ground as the year goes by and a simple evaluation of numbers based upon it. Most deer managers rely very heavily on this process, although of course it is a subjective method based on experience and intuition. Someone new to an area simply does not have enough accumulated knowledge to carry it out but on smaller pieces of ground with limited deer populations, informal observation may be sufficient to provide a simple 'rule of thumb' check on numbers.

The rolling census may not be a formal approach but it has value. It is sometimes referred to as an 'activity index'. The better you know your ground and the deer that it contains, the better the impression you will gain of overall numbers – as well as being able to recognise unusual increases, or indeed population decreases. Times of year can be deceptive to the uninitiated. There may be seasonal migrations, caused by variations in food sources, weather or more specific disturbances, and sometimes the deer can seem to disappear entirely. Roe are particularly notable in this respect, going through a period of seasonal inappetence at the end of the year when, to conserve energy, they move and feed very little, and simply do not show themselves. This phenomenon seems to vary from location to location but on one piece of ground that I came to know very well in Hampshire the roe were virtually invisible from the opening of the doe season in November until just after Christmas, when they suddenly started to reappear in good numbers.

There is nothing wrong with relying on a rolling census, but for management purposes it is best to back up informal observations with one or more formal methods at regular intervals.

Direct counts: vantage point

The ways of counting deer can be split into two distinct approaches: direct counts, which rely on seeing the deer themselves, or indirect counts, which look for the signs of their presence and evaluating these without any need actually to see the deer themselves.

↟ *Under open landscape conditions it is easier to count deer while assessing their sexes and ages.*

Vantage point counts can produce some very detailed and accurate results. They do, however, require open ground with long lines of sight and lend themselves best to moor or heathland. The method is simple enough: observers are positioned to cover as much of the area in question by sight and, armed with binoculars, record all deer seen throughout a set period. Radio communication between observers will help to prevent double-counting. On suitably open ground the deer may be observed even if not on the move, so the count can take place throughout much of the day; but if there is any amount of cover available for the deer to lie up in, the timings may need to be considered more carefully, or arrangements made to move the deer so that they can be seen.

The observers themselves need to be relatively experienced if anything beyond a basic count is required, but otherwise a good idea of male to female ratios, and even age classes, can be obtained. The manpower bill can be high if a large area is to be covered effectively, although this will be decided by local topography. Longer views usually mean that fewer observers will be required. Local co-operation between landowners can make it possible to cover neighbouring patches on successive days.

There are a few disadvantages to this method, and of course the weather can have a major effect on its success. Likewise, it is important to remember that some deer species tend to move seasonally, so numbers and sex and age mixes might vary according to the time of year. On the plus side, the method causes minimum disturbance to the deer, beyond perhaps a need to use walkers to push out small areas of cover.

Direct counts: observation point

Observation point counts are similar in many ways to the vantage point type, but are better suited to more enclosed areas such as woodland and associated farmland. Because of the closer cover, and therefore limited areas of observation, it is important to accept that a lesser degree of accuracy is likely to be obtained. Once again, observers are positioned to cover as much of the ground as possible: but it is unlikely that they will have lines of sight to each other, even in fairly sparse woodland. As this method is more reliant on the deer showing themselves, the count needs to be carried out at times when the deer are more likely to be on the move, namely the couple of hours after dawn and before dusk respectively.

A simple observation point count. The red triangles represent observers sited to cover rides, wood edges, and more open areas.

Manpower requirements are proportionately higher for this method, and observers need to be sited carefully. Once again, co-operation between neighbours over successive days can be helpful. Both permanent and portable high seats are useful fixed points from which to watch. There is a very great danger of double counting; radios are less effective and can of course disturb the deer at close quarters, so it is helpful to issue observers with cards on which they can record sightings. These cards need to show species, sex, time, location and direction of travel. By collecting these in and comparing them at the end of the count, a clearer idea can be obtained of which deer, or groups of deer, may have been seen at different times from different points.

Observer: A J M
Location: BOX SEAT @ TRACK JN Date: 8 MAR 18
GR: SU 123456

Time	Species	M/F	Unknown	Direction	Comments
1610	ROE	1 x f	1	W	DOE AND KID
1615	ROE	1 x m	—	SW	MATURE BUCK
1705	FALLOW	2 x m 6 x f	2	W	PRICKETS + DOES 2 CALVES.
1720	MUNTJAC	1	—	↗	most common 1 BLACK DOE + 1 MENIL DOE
1740	ROE	1 x F	—	↖	GOOD BUCK
1745	MUNTJAC	1 x F	1	W E	YEARLING ? DOE + FAWN

Card 1 of 2

A record card for observers. It should be produced in a dull colour so that movement is less likely to alert deer in the vicinity.

As with vantage point counts, the observers all need to be in position in good time and to know when the finish time is. A briefing beforehand is essential to ensure that all are aware of any administrative points and, more importantly, emergency procedures.

Direct counts: thermal imaging

The advent of high quality, affordable thermal imaging equipment (see Chapter 5) has provided an important tool for the deer manager and offers some major advantages. Firstly, it does not require large amounts of manpower to use it when conducting a count. Even for larger pieces of land an observer with the TI equipment and a driver are quite sufficient, although a third person to take notes can save a great deal of time. Secondly, TI enables sizeable areas to be surveyed in a session, and even moderate amounts of ground cover do not restrict what can be picked up. For species such as fallow, which can become almost entirely nocturnal when subjected to intensive management, it is probably the only way to get a fair idea of the actual numbers present.

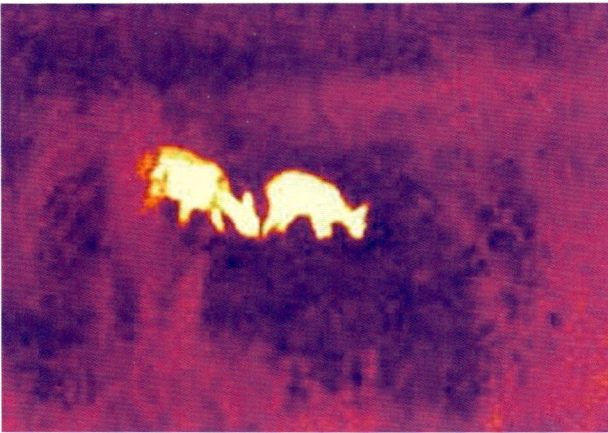

↑ *Thermal imaging can be effective, although precise identification of species, sexes and age groups may not always be possible.*

This method does have some disadvantages though. The cost of equipment can preclude it in the first place, so it is not something that is available to everyone. While compact models can be picked up for a few hundred pounds, these tend to be of a lower quality and have shorter effective ranges. Top-end models do not come cheap and it is possible to pay several thousand pounds for something really practicable. Furthermore, the chances of accurately identifying the sexes and ages of deer are reduced as the distances from them increase, and it follows that equipment operators need a degree of training and experience in its use.

A word of warning may also be appropriate to anyone considering a cut-price model, as some are still occasionally offered as military surplus from around the world. Not only can sources of repair and servicing be limited, but they may also rely on power sources not readily available in this country. Modern commercial models, on the other hand, are covered by guarantee and technical support, and a higher initial outlay tends to be a saving in the long term.

Direct counts: spotlighting

Spotlighting provides another opportunity for counting deer at night, and the equipment itself is much cheaper than the thermal imaging type. It is most effective when deer are in the open, and of course white light cannot 'see into' cover in the way that devices that register heat can. Nevertheless, some accurate counts can be achieved if the equipment is appropriate for the task and it is used in combination with binoculars.

The method is better for species which customarily feed in the open at night, though it is potentially much more disturbing to the deer. In places where there are issues with poaching, animals will probably be less tolerant of being illuminated and may well have learned not to stand for long enough to be counted or accurately identified.

Direct counts: moved counts

The moved count can be very effective, but can frequently be dependent on having the right circumstances. In this method, the deer are encouraged to move quietly away along their accustomed routes by gentle human disturbance – they are certainly not driven in the manner of a game shoot. Observers, carefully placed, can then record them.

Where deer are liable to have a large number of exit points from which they can depart unobserved, the moved count can be ineffective, but in large, semi-enclosed areas or woodlands it can be a practical option. In addition to observers, who need to be experienced and capable of making a quick and accurate identification, a line of walkers is also required to ensure that the deer keep moving forward. Dogs are better left at home unless they are extremely steady and reliable, although they can be useful for investigating thick patches of cover such as bramble beds where humans are loath to go. The moving line needs to be dense enough to ensure that deer cannot simply lie up and watch people go by, and manpower requirements will vary according to the terrain and cover available.

If the ground allows it, a move can be taken in several manageable stages, using major woodland rides, cleared ground or patches of woodland as distinct areas to be worked through before static observers can be repositioned and the walking line moves on. Any deer that breaks back through the walking line can be recorded and then reasonably discounted from the remainder of the move.

Sometimes the deer will simply not co-operate, as the example that opens this chapter illustrates. Overall, the method seems to lend itself best to roe, as other species, especially muntjac and fallow, can resist being moved into sight and may simply double-back and hide in thick cover. Double counting can be a major problem, and protocols need to be put in place to prevent this as far as possible: for example, deer breaking back through the line of walkers should only be recorded by the person who sees the animal passing to their left-hand side. Static observers, where possible, should have a line of sight to each other and observe similar rules. Everyone taking part needs to be issued with a simple, locally-produced card for recording their sightings. These can be collected and the information consolidated at the end of the process.

A major disadvantage of the moved count method can be the manpower required. To cover an area some 500 metres wide you may need as many as 40 or 50 people to stay in contact across the walking line where the cover is dense, with a further dozen or more static observers positioned on firebreaks or wood edges where they can see each other over longer distances. A comprehensive briefing beforehand is of course essential, but the difficulty of maintaining control over an extended line of walkers, all of whom need to keep in a straight line and remain in sight of each other at all times, cannot be underestimated.

⬇ High visibility jackets help all those involved in a walking line to stay in sight of each other, especially when the cover is close.

Direct counts: other methods

While some may suggest that an **aerial census** by helicopter is an ideal way to conduct a count – and there is no doubt that

↑ *Observation from the air, either by drone or helicopter, can be effective in open areas and may be enhanced by the use of thermal technology.*

large areas of open ground can be covered quickly and effectively – the sheer cost means that this is not a practical option for the average landowner or deer manager. Nevertheless, helicopter counts are sometimes carried out, albeit usually by government organisations, so the method must be included as an option.

The growth in the availability of **drones**, with a live camera-feed to either tablet or computer, presents a very real opportunity more likely to be within financial reach, and improved battery capabilities are making them ever more practical for extended surveys. Thus it is increasingly possible to use them to cover large areas of ground while recording the results for future analysis. However, their use demands only the patchiest of ground cover at most, if the results are to be successful.

➡ *Drones are becoming an increasingly realistic and affordable option for aerial surveys.*

Photograph: DJI

Drones are becoming increasingly popular amongst hunters, particularly in the USA, for 'scouting' areas for the presence of deer, and recently a project has been announced in Japan to survey areas of forest using drones fitted with thermal cameras. A linked computer programme will analyse the collected data against a database of images to distinguish deer from other animals.[6] Drone users do need to take care to observe the law and best practice as, if inappropriately used, the devices can have an adverse effect on wildlife and even result in the user being prosecuted. The Civil Aviation Authority and the Partnership for Action Against Wildlife Crime in Scotland provide some very sensible, good practice advice, available online.[7, 8]

Trail cameras, sadly, have negligible usefulness as far as counting deer is concerned. Being set in fixed positions with limited triggering ranges, the sheer number that might be required to cover effectively any but the smallest pieces of ground makes their use impractical, even if it is possible to time- and date-stamp images to avoid double counting.

Road traffic accidents, however, can be a useful indicator of population growth or decline provided that accurate records are kept, as can **anecdote** which should never be disregarded. Taking the time to speak to other land-users may provide valuable information regarding the whereabouts of deer and their movements as well as a historical background. A talk with foresters and farm workers, for instance, who have worked an area for many years, will often provide a wider picture of trends which might not otherwise have been apparent to a relative newcomer.

↓ *Dung counting demands a meticulous approach and can be very time-consuming.*

Indirect counts: dung counting

Probably the best known, if not properly understood, indirect method of deer population assessment is the dung or faecal pellet count. Once popular, not many people outside the academic community tend to employ it these days. Whilst having value in dense forestry situations and potentially being very cost-effective over larger areas of woodland, the technique is not popular among most deer managers.

In simple terms, there are two main approaches. In the 'faecal standing crop' method, the dung is simply counted within a specified plot, taking decay rates into account. This has the advantage of being possible to complete in a single visit.

More regularly used is the 'faecal accumulation rate' method, relying on a number of specific plots located in different habitat types. Plots either measure several metres square in typical cases, or may be linear transects of perhaps 1 metre by 50 metres that stretch completely across the habitat type. They are marked by posts from which a temporary tape can be stretched to mark them whilst the work is being completed, then left in place until the second survey visit. These plots are cleared entirely of any existing deer pellets, leaving one (marked) fresh set in each so that decay rates can be assessed.

At a later date (typically two to three months later, but taking into account the potential for decay depending on habitat, weather conditions and so on), the plots are revisited and once again the dung is counted and assessed. A formula is then applied which involves the size of the study site, the number of dung groups found,

6 Nikkei Asian Review (2016), *Report: Chinese drones to count Japanese deer*, 25 June 2016

7 Civil Aviation Authority (2015), *Unmanned aircraft and drones*; available from www.caa.co.uk/ Consumers/Unmanned-aircraft-and-drones (Accessed 2 September 2018)

8 Partnership for Action Against Wildlife Crime in Scotland (2018), *Good Practice Advice – Drones and Wildlife*; available from www. gov.scot/Topics/Environment/ Wildlife-Habitats/paw-scotland/ Resources/Goodpracticeadvice/ drones (Accessed 2 September 2018)

the time involved in their accumulation, and the defecation rate of the specific deer species. Through these an estimate of the number of deer using the various habitats can be calculated.

Whilst a relatively straightforward process and not an expensive one in terms of equipment or needing large amounts of manpower, it is easy to see that dung counting is very time-consuming and that there is a great deal of room for error if it is not done carefully. The counter also needs to be experienced in differentiating between the droppings of different species of deer. Of necessity having to be completed at times of the year when vegetation is low so that pellets can easily be found, a dung count can be further complicated by unexpected weather conditions such as a heavy fall of snow.

Should you wish to try this approach, there is a thorough description of it, as well as other deer census techniques, in the Forestry Commission handbook *How Many Deer* or the Deer Initiative Best Practice Guide on the subject.[9, 10] However, for mainstream deer management there are probably better and less complicated methods to employ.

Indirect counts: environmental surveys and impact assessments

Some deer, such as muntjac, are reluctant to leave cover as already noted and offer few opportunities to be counted as individual animals. In such cases it is best to avoid assessing the deer themselves directly, but instead to look at the impact that they are having on the environment. By this means a working idea of the population can then be achieved. The result manifests itself not in actual numbers of deer, but rather as an assessment of their density in terms of low, medium or high and a recognition of the effect that this is having on the landscape. An experienced eye is needed, for the observer needs to be able to interpret the signs of deer presence such as those they leave by their day-to-day activities.

A visit to some ground in Oxfordshire a few years ago to assess muntjac numbers illustrates the value of this technique. In this particular area, disagreement over deer numbers had led to a questioning of management policy as very few were actually being seen. In fact, a day of carefully surveying the woods for tracks, fraying, droppings and other signs led to the conclusion that the muntjac population was not only high but, taking historical records into account, actually increasing – with potentially disastrous consequences for the other flora and fauna in the woods. Throughout the day only two muntjac were physically seen and only two sets of roe tracks had been noted. This in itself was significant, given that that particular set of woods had once held a strong roe population which, it was surmised, had gradually decamped as muntjac numbers had grown. A reinstatement of muntjac culling subsequently resulted in the re-establishment of the woodland understorey, an increase in the recorded ground-nesting bird species, and the return of roe deer to the woods.

If this approach appears somewhat informal, more structured impact assessment methodologies have been developed to take matters to a rather more objective level. Initially based on a simple scoring system increasingly popular among conservationists and others, these look at vegetation affected by grazing, browsing and fraying.[11] Deer sightings, slots (tracks), droppings and paths are also

9 The Deer Initiative (2008) *Best Practice Guide – Dung Counting*; available from www.thedeerinitiative.co.uk/uploads/guides/175.pdf (Accessed 29 September 2018)

10 Mayle, B.A, Peace, A.J. and Gill, R.M.A. (1999), *How Many Deer? A Field Guide to Estimating Deer Population Size*, Forestry Commission Field Book 18; available from www.forestry.gov.uk/PDF/FCFB018.pdf/$FILE/FCFB018.pdf (Accessed 22 February 2018)

11 Cooke, A. (2006), *Monitoring muntjac deer* Muntiacus reevesi *and their impacts in Monks Wood National Nature Reserve*, English Nature Research Report No. 681; available from www.publications.naturalengland.org.uk/publication/210342 (Accessed 29 September 2018)

← Evidence of fallow browsing on an ivy-covered tree trunk. Impact assessment looks at a variety of signs of deer presence.

assessed. Scores are allotted to each aspect ranging from 0 (low) to 3 (high) and ultimately a benchmark total score is arrived at. Over time, this provides an annual comparison of scores and the results determine whether deer impact is reducing or increasing, and culling efforts adjusted accordingly. Future annual assessments and the scores achieved indicate the success (or otherwise) of any remedial plans.

To avoid seasonal variations, the assessor should take care to walk the same defined routes (usually trackways) at the same time of year. Results are of course very subjective, and it is important that the same assessor completes the survey on each occasion to ensure that they are consistent. Though remaining somewhat crude, the system has proven very successful for the management of muntjac, and has since been adapted for use with other species of deer. It can be further amended to provide scores for impacts on specific plant species, such as orchids or bluebells, which might be vulnerable to grazing pressure; and it can also be tailored to an individual wood and all that is known about it.[12]

Where more than one species of deer is to be found in a given area, it can of course be difficult to differentiate between some of the signs left by the different species, and a considerable degree of expertise is essential. ■

12 Cooke, A. (2019) *Muntjac and water deer: Natural history, environmental impact and management.* Pelagic Publishing, Exeter

7 Cull Planning

If you have considered the non-lethal alternatives and decided that a reduction of deer numbers is necessary, it is worth taking a moment to consider the meaning of the word 'cull'. The Oxford English Dictionary defines this as: 'the act or product of culling; a selection. An animal drafted from the flock as being inferior or too old for breeding'. It is, just as the definition explicitly suggests, a selective rather than haphazard process if carried out correctly, and should result in a better, more balanced population. Conducted thoughtlessly, though, it can create imbalances and do more harm than good.

I have always wondered if there is a direct correlation between stalker efficiency, deer density and time spent on the ground, which can be wrapped into a neat little formula that might do away with the need for formal cull planning. Sadly, though, there are so many variables woven into these factors that such an academic exercise is rendered almost impossible. Nevertheless, for many sites it is feasible to come pretty close to it. Smaller areas can lend themselves to the deployment of a small number of stalkers, acting with restraint, to whittle the deer population down to a level perceived by landowners and managers to be acceptable.

Restraint, though, must be the watchword: one undisciplined or greedy hunter can do immeasurable damage to herd structures, so this more informal, unplanned approach is probably better restricted to smaller management areas where big team efforts are not necessary. If you do decide to control numbers in this way, try to direct the bulk of your efforts towards the younger male animals and females.

For most areas, though, and especially where larger sites and cross-border co-operation are involved, it is essential to take a rather more formal approach.

Selective control

As you begin to formulate any management plan where deer are concerned, it is important to accept that some degree of damage by the animals is almost inevitable. Selective control will, however, help to keep this damage in proportion. It will take advantage of the natural characteristics and behaviour of each species in question and may aim to limit the population, as well as taking advantage of natural territorial aggression by encouraging the presence of more dominant animals. With roe, for example, it is far better to accept the relatively light territorial marking of a few strong, middle-aged 'stand' bucks as opposed to the disproportionately higher damage inflicted by a multitude of bickering youngsters.

← *Restraint is often necessary when faced with culling choices.*

➜ *A strong, middle-aged roe buck will hold a larger territory, effectively protecting it against the depredations of higher numbers of younger animals.*

Balancing sex ratios reduces fraying damage as there is less competition. Similarly, balancing the overall herd against the carrying capacity of the available habitat means that it is better able to exist comfortably on the natural resources that are there. With larger deer species, this will manifest itself in less winter-bark stripping, for example. An intelligent deer manager, knowing the herd structure and habits (especially seasonal movements) of the resident deer, will come to terms far more easily with their management objectives.

Selective control, apart from minimising damage, has the added benefit of increasing both body weights and antler quality, encouraging a tangible benefit in terms of venison yields and stalking incomes. This in itself can offset any costs of the damage caused by the deer in the first place. By managing deer as part of a forestry or agricultural harvest, the animals can coexist comfortably alongside other interests as another renewable resource.

Management objectives

Before you start to plan your cull it is important to decide exactly what you are trying to achieve, as all of your subsequent calculations will depend upon this. Sometimes such decisions may be out of your hands if the landowner has fixed ideas, although you may of course be in a good position to advise and influence. If a deer management group, comprising of different landowners and interested parties over a wider area is involved, the matter may become somewhat more complicated but nevertheless a compromise needs to be reached.

Occasionally it just has to be accepted that pure management objectives may not be achievable. Some years ago I managed a piece of roe ground in the south of England where, despite all my best intentions, it was impossible to foster any quality amongst the deer. The reason was easy to see. The various neighbours included a farmer who shot anything as seen 'for the pot', a sporting estate that consistently oversold trophy shooting to clients, and a nature reserve where no controls were applied at all. Not only that, nobody spoke to each other. As a result, the deer population on my ground was comprised mainly of does and small bucks, which caused disproportionate damage in the few small plantations the local forester was trying to establish. It was extremely frustrating, and the best we could do was to try to keep the deer numbers down to a moderate level and maintain healthy sex ratios, while shooting as many of the yearling bucks responsible for the excessive forestry damage as possible.

Four essential truths of deer management:

1. **It is virtually impossible to eradicate deer.** *Even if you do, you are probably just creating a vacuum that others will quickly fill.*

2. **To control numbers, stags and bucks are almost irrelevant to the long-term plan.** *It is the females that produce more deer. A group composed entirely of male animals will eventually die out, whereas a healthy female population will continue to reproduce even if there are only a few males in the vicinity.*

3. **Mistakes happen.** *They are generally not catastrophic in the larger scheme of things. Learn from them, and try not to repeat them!*

4. **Flexibility is vital.** *Deer management is not an exact science. Be prepared to react to changing circumstances and do not be afraid to adjust or change your plans if necessary.*

We need to be clear from the start: under most circumstances, simply **eradicating** the deer is not an option. It has been achieved in some places, but only with the assistance of massive manpower resources and then considerable expenditure on fencing to prevent repopulation after the existing one has been removed. Even then, deer populations put under intense pressure may simply change their habits and become extremely difficult to manage, in some cases moving and feeding entirely during the hours of darkness. Frankly, I would discount eradication as a realistic objective for any but the most unusual of situations.

Where deer numbers are too high, it is more sensible to embark on a **reduction cull** to bring them down to an acceptable level. It is essential to direct your efforts towards the female side of the population, as these are the animals that physically produce more deer. With the exception of muntjac, which have no close season, it follows that your busiest time is going to be throughout the winter months, a period complicated by short days, inclement weather and, where game shooting has priority, limited access to the woods for much of the doe season. Even for muntjac, which may legally be shot all year round, new growth in the woods may well render them virtually invisible for much of the spring, summer and early autumn, so even for this species winter will be your time of maximum effort. Stags and bucks of the larger species may be in season at the same time, but where opportunities occur the stalker should always take the doe or hind if possible. This may, at times, demand a very great deal of self-discipline.

↑ In this area of Caledonian pine, a reduction cull has allowed regeneration to take place where it was previously impossible due to excessive deer numbers.

Throughout the process, it is advisable to remain aware of the effect that your efforts are having on herd dynamics; as I have said, deer put under too much pressure may alter their habits and become very difficult to see and to deal with. Do not expect to achieve your reduction cull overnight – in some cases it might take many years of hard effort.

Once you have reduced numbers to a sensible level, you can implement a **maintenance cull** to keep them that way, in line with your overall objectives. To keep population sizes stable, and provided that the sex ratios are balanced, it will probably be necessary to cull around 20% of the population of the larger species (red, sika and fallow) annually, whereas for the smaller species (roe, muntjac and Chinese water deer) some 30% will usually be required. At this point you can start refining your approach towards more specific aims.

The maintenance cull allows you to do this fine-tuning. Perhaps the purest approach to deer management is simply to keep numbers balanced against the environment which supports them, and all that needs to be done is to determine the holding capacity of the ground and reduce the deer numbers accordingly. Although that sounds very simple, it is important to pay due consideration to sex ratios and age structures as we have seen; but essentially what you are conducting is a natural cull. Beware, though, that you need to leave a margin for error in case of severe weather and reduced food sources – but equally the chances are that if you maintain numbers at too high a level the overall condition of the animals may be affected. As ever, it is necessary to consider your neighbours too, who may experience higher levels of agricultural or forestry damage than they find acceptable, and road traffic accidents may also be an issue in places. Inevitably, by keeping numbers at a higher level you are permitting the potential for more rapid population expansion if care is not taken.

Otherwise, your management objectives will now determine the level at which you hold numbers. A desire to maximise venison revenue demands a greater number of deer (especially females, to produce more deer in their turn) on the ground for harvesting, but this may result in greater effects and impacts elsewhere. Conversely, a hoped-for improvement in trophy quality usually requires numbers to be held at lower levels, to reduce stress and disturbance and also to enhance the habitat and forage available to allow the remaining animals to produce better antlers.

If there is a local desire to minimise deer impacts, you may need to reduce numbers to well below the actual carrying capacity – this could indeed be an imperative when agricultural and other similar interests are to the fore. In essence, though, the chances are that you will be conducting a delicate balancing act between the requirements of landowners (who may actually have a mixture of objectives in mind) whilst simultaneously making a personal effort to improve overall herd quality and taking an ethical approach towards deer welfare.

↑ *Most balanced culls will always focus on females and young animals of both sexes.*

Carrying capacity

The carrying capacity of a piece of ground, namely the number of deer that it is capable of sustaining naturally, will vary enormously depending on any number of factors. Not least is the **food** available, and even then this will be subject to the size and dietary requirements of the deer species that rely upon it. Small deer species, such as roe and muntjac, have correspondingly small digestive systems and as 'concentrate selectors' need smaller amounts of higher quality fodder. A bigger red deer, on the other hand, may be better able to process coarser foodstuffs but will require them in larger quantities. Add a mixture of deer species to the equation and matters can get even more complicated. Overall, the availability of suitable food will have far-reaching effects on condition, births, deaths, emigration and immigration across a deer population. Food availability must always be considered both at times of plenty but also when it is reduced according to seasonal growth or harsh weather conditions.

For many deer species, the **cover** available to them will also have a major effect on the numbers that the ground can support. When a plantation is at its thicket stage, there may be plenty of shelter but little food; once thinning has taken place and the tree canopy begins to open up, food availability will increase. Deer can of course live in the open on moorland or grassland, where only the topography of the ground and some restricted scattered cover allows for a degree of protection from the elements, but only much lower densities can generally be supported by such habitat.

↑ *Carrying capacity must take account of those times when forage and available cover are likely to be least available.*

Social habits will also play a part. Herding species may be tolerant of each other in larger numbers but roe in particular tend to be solitary for much of the year and can be deeply intolerant of others of their own species. However, such territorial aggression usually tends to be seasonal.

Human activities and **land use** will have a profound effect on deer populations too. For example, the clear felling of a large area of woodland or the sudden introduction of domestic stock can reduce the useable habitat almost overnight and may demand a rapid and realistic re-evaluation of just how many deer the area can now reasonably support. Similarly, recreational disturbance - such as orienteering events and in particular mountain biking - or an increase in the establishment of public access areas, may have long- or short-term impacts.

It is impossible to specify hard and fast rules for carrying capacities. In his definitive work, *The Roe Deer*, Richard Prior observes that in one place that was intensively coppiced, the roe population probably exceeded one animal for every 3.5 hectares, while elsewhere it might take 33 hectares of mature conifer plantation to support a similar number.[1] Scottish Natural Heritage note that red deer densities can vary wildly from 1 to 20 animals per square kilometre (i.e. from 1 per 5 hectares, to 1 per 100 hectares), and that conditions such as average snow cover have a major effect.[2] Interestingly, the same source notes that the presence of sheep may cause hind, but not stag, numbers to decline. Where conditions are ideal, muntjac have

Deer Population and Carrying Capacity Relationship

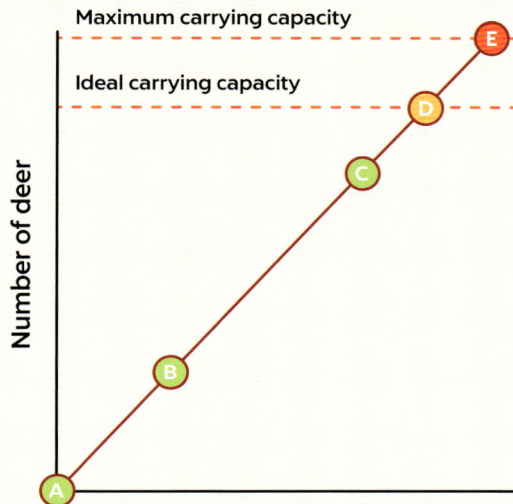

Maximum carrying capacity

Ideal carrying capacity

Number of deer

At Point A the deer population is minimal to non-existent, although Point B probably represents the lowest density to which it can be realistically reduced with normal resources and without causing excessive stress; overall herd condition, body weights and antler growth will be optimised.

Point C maintains the population safely within carrying capacity. Venison yields are high although impacts on the available habitat are becoming more evident.

At Point D body weights may be showing signs of reduction and external parasite counts are rising. Territorial stresses may be starting to show.

When the population reaches Point E, habitat is being very noticeably affected and birth rates may be reducing. More animals will be observed in poor condition and the potential for disease is increased. Any failures to meet cull targets are likely to reflect in increased mortality, especially during the winter.

been recorded in densities as high as 150 per square kilometre of woodland, a staggering 1.5 animals per hectare.[3] Roe density in southern England is commonly in the region of 50 per square kilometre of woodland.

Assessing the ideal level is never simple, and will be complicated even further when you have more than one species of deer on your ground. There may well be competition between the different species for available food and cover, although some seem to mix better than others. Roe and muntjac, for example, being selective feeders, can each deplete the resources that the other needs; this must be taken into careful account when planning. The larger deer species, on the other hand, are less likely to rely on precisely the same food plants. Nevertheless, where there is a mixture of species, the overall carrying capacity of the ground will inevitably be affected and you will need to reduce your estimates accordingly.

It is obvious that carrying capacities vary enormously from place to place and according to different circumstances, and that it is impossible to lay down absolute rules. Only the person with relevant knowledge, experience of the ground and an observant eye is in a position to make a judgement on what is suitable for a specific piece of ground. Even then, this judgement must be considered subjective and a degree of flexibility is always necessary.

The Cull Plan – Factors Involved

Assuming that you have taken a census and have a figure for the deer on your ground that you are reasonably confident is fairly accurate, you are now in a position to start planning the actual cull for the year ahead. However, it would be a huge mistake to believe blindly that a census figure arrived at in March will necessarily forecast the number of deer that will be on your ground throughout the

1 Prior, R. (1995), *The Roe Deer*, Shrewsbury, Swan Hill Press

2 Scottish Natural Heritage, *Variation in red deer density in the Highlands*, Information and Advisory Note No. 100; available from www.snh.org.uk/publications/on-line/advisorynotes/100/ (Accessed 26 February 2018)

3 Griffith, D. (2004), *Deer Management – Quality in Southern England*, Fordingbridge, Dominic Griffith

year to come. There are a number of variables to be considered in your forecasting. These variables, by the way, are not just restricted to deer and apply to pretty well any form of wildlife that needs managing: they are really very straightforward and once understood become quite intuitive to apply.

Fecundity

The term 'fecundity' simply describes the rate at which a particular species is capable of breeding, or to put it differently, 'a measure of the capacity to produce offspring'. It is not the same thing as fertility, which defines an ability to breed and no more. Furthermore, whether all the young actually survive after birthing is not entirely relevant to our consideration of fecundity, so mortality will be looked at separately.

For the purposes of management planning, we tend to take fecundity as an average figure that may vary from year to year. This estimate is of huge importance to your plans and needs to be examined in detail. Variations may result from a number of causes, not least of which are the conditions that the deer live under. At times of hardship (such as a lack of food) fecundity has been noted as declining. While red deer normally reproduce year on year, it is a well-known phenomenon that red deer hinds living in the harsher conditions of the Scottish Highlands will commonly miss a year of breeding if they are of insufficient body condition at the rut, whereas lowland red deer in a kinder, semi-woodland environment may have double the calving rate of hill beasts.[4] Thus the fecundity rates within the same species can vary wildly across different parts of the country.

Roe deer are another good example. Typically, a roe doe will produce twins throughout her breeding life but, where conditions are especially good, triplets are occasionally produced – though this is heavily dependent on the resources available to the animal during her pregnancy. In most cases, fecundity and condition go hand-in-hand.

➡ *A roe doe will typically produce twins at each birthing. Of the other British species, only the Chinese water deer habitually produces more than one offspring.*

Not all of the female deer in a population will breed even under the best conditions. A proportion may be younger animals which are not yet fertile, while some attain an age where fertility eventually decreases and conception does not necessarily occur during the rut. The different species may also breed at different ages. Sexual maturity is usually linked to the young animal attaining a threshold body weight which enables it to start to ovulate or, in the case or males, to produce sperm. The smaller species tend to reach this threshold weight more quickly because less physical growth is required to attain it.

All species, with the exception of the muntjac, are further constrained by fixed times of the year for rutting activity. For example, a roe doe born in May of Year 1 will not be sexually mature in time for the rut in July or August, so will not mate until Year 2 and then, thanks to the mechanism of delayed implantation, not give birth to her first kid until her second birthday in Year 3. The larger deer species, although having similar birthing times but a much later rutting period, still cannot attain the rapid growth rates required to take part in the latter although, very occasionally, a particularly precocious calf assisted by exceptionally good conditions may manage it. Such instances, though, tend to be extremely rare.

The greatest potential for rapid reproduction lies with the Chinese water deer and muntjac. The Chinese water deer ruts very late in the year (a legacy from their Asian origins) and it is not unusual for a doe born in May or June to be sexually mature in time for the December rut and then give birth on her first birthday. As multiple birthers (as many as six fawns have been recorded, although two or three are more usual), their potential fecundity is very high. The muntjac, on the other hand, our one deer that has no fixed time for rutting and birthing, can give birth at any time of the year. Both bucks and does are sexually mature at about seven months of age, and gestation is another seven months, so a doe can give birth for the first time at around fourteen months. Fecundity is lower though, as muntjac produce only one fawn at a time - although twin foetuses are occasionally recorded.[5] Although does have been seen accompanied by two fawns of about the same age, there is currently no hard evidence available to prove twin births. Subsequent mortality caused by bad weather, high predation and other adverse conditions is of course another matter.

Some suggested annual fecundity rates for deer populations of different species are listed below. These can never be taken as hard and fast rules, and as we have seen they will need careful adjustment to take local factors into account:

Muntjac	1.7
CWD	1.8
Roe	1.6
Fallow	0.9
Sika	0.9
Red (lowland)	0.8
Red (Highland)	0.5

4 Alcock, I. (1996), *Deer – A Personal View*, Shrewsbury, Swan Hill Press

5 Chapman, N. (1991), *Deer*, London, Whittet Books

↑ The large deer species usually give birth to just one offspring at a time, although inexperienced observers may form a false impression. This sika hind is tolerating suckling by a number of calves that are not all her own.

The male-to-female ratio in the adult deer population is not vitally important under any but the most extreme circumstances. Among the larger herding species, the majority of females will be served by a small number of dominant stags or bucks, while in the smaller, more territorial species it is the doe who tends to seek out a buck to rut with. Although some deer, such as roe and muntjac, may give the impression of forming close relationships, they tend instead to maintain overlapping territories and home ranges which ensure that they are in close proximity as the rut approaches. No deer are truly monogamous, and a male is capable of serving any number of females in oestrus.

Whilst the closed seasons for shooting female deer are deliberately set so that dependent young are not inadvertently orphaned, signs of pregnancy in culled animals are a useful guide to the general prospects for birthing the following spring. Even for roe, with their long period of delayed implantation when the fertilised egg lies effectively dormant within the uterus, it is possible to recognise the number of fertilised eggs present by looking for the *corpora lutea* as part of the gralloch. These are large yellow glandular bodies in the ovaries which are easily recognisable with practice. Keeping records comes into its own here, and trends identified in previous years can be compared to those being currently seen.

As we have noted, fecundity rates are very different for the different species and even then they may vary from year to year. Local knowledge is invaluable when forecasting how many young your local deer are likely to produce, and future fecundity trends have to be evaluated carefully when calculating a cull plan. However, it must also be stressed that, as for most matters concerning deer, such forecasting must be treated not as an exact science but rather as educated guesswork!

Mortality

In this context, mortality is the term used to encompass all the factors causing deaths within a given population of deer. Natural mortality of deer in the UK is generally the result of malnutrition, weather, parasites and disease, old age, or a combination of all these factors. Elsewhere in the world predators can be added to this list, and indeed predation can be a very significant aspect of deer mortality in many places. By the time a deer is debilitated for other reasons, it inevitably becomes easier prey for carnivores than its fellows, and is culled via the process of natural selection. Here in the UK deer have no natural large predators, so instead the human replaces them, with the exception of young calves or fawns that are vulnerable to the badgers, foxes and others which might predate upon them. The smaller fawns of the muntjac and Chinese water deer, being little larger than rabbits, are also susceptible to predators as small as stoats, and of course domestic dogs and cats are a threat to the young of all species. Dogs are perhaps the only mammals in the UK that are capable of posing a threat to a healthy, mature deer of any species.

Under ideal living conditions with adequate food, clement weather and an unstressed population, old age is often suggested as being the major natural cause of mortality among deer. In fact this is not true. The adult teeth of a deer are not replaced as they wear down, and eventually an animal will reach the point where it cannot process cellulose sufficiently for the digestive processes to extract full nutritional value from it. The result is then death from malnutrition: this is not starvation, as the unfortunate animal may well have a full stomach at the time of its demise.

British deer living at the extreme edges of their ranges, such as in the Scottish Highlands, are particularly vulnerable to severe weather and the poor feeding especially associated with the times when it tends to occur. Furthermore, spring mortality can be high if the already weakened deer are subjected to heavy rain accompanied by chilling winds. Hard weather is very likely to have a significant effect on young deer entering their first winter; not being fully grown, and with the resources they consume being put into growth rather than laying down fat reserves, they have less to fall back upon to sustain them during times of need.

Deer, and roe in particular, have a coping measure designed to help them through times of low food availability, that of winter inappetence. At the start of the roe doe season in November, it often seems that the deer have all but disappeared: whereas in fact their metabolisms have simply slowed with a correspondingly reduced need for food and a tendency to move about much less. Once Christmas has passed the deer generally become more active and visible again.

⬆ An efficient deer manager will probably come to know individual animals within a population and be able to judge when they have started to decline and cull them before natural mortality occurs.

⬇ Adverse weather at birthing times, and for the weeks that follow, can be a major cause of fawn mortality.

Apart from the weather and other factors, young deer face particular dangers and may suffer mortality during their first few months of life from rejection by their mother, accidents, predation or dehydration, the latter occurring during unusually dry conditions when the mother finds herself incapable of producing sufficient milk for successful suckling. Extensive studies of red deer on the Scottish island of Rhum have shown that some 10% of calves die during their first two weeks of life.[6] As many as 20% of calves born in June might not survive until the end of September, with a further 11% of the survivors not making it through the winter. In the case of Highland red deer, reliance on the mother can last for a surprisingly long time when compared to other species; should she be lost, this can affect the survival rates of young deer of both sexes, and the antler growth of yearling males, for as long as two years after birth.[7] The strain on the mother herself of raising her offspring cannot be discounted either, and the burden of rearing dependents may have an adverse effect on her own ability to maintain condition in a difficult climatic environment.

↓ *Red deer calves may remain dependent on maternal care for much longer than other species.*

Mortality amongst fallow is also significantly high during the first year of life and this has been noted even under park conditions where winter takes its toll of body condition, with deaths following in the spring. As for all deer, the loss of a mother and the lessons in survival that would otherwise be learned from her by a young animal can be catastrophic even very late in the year. In a Richmond Park study the average weight across 24 fallow fawns that died in February and March was only 18kg (ranging from 13.6 to 29.5kg), against the average of 27kg (ranging from 20 to 34kg) for another 31 fawns deliberately culled during the previous November and December.[8]

Even with the advantage of multiple births, mortality among roe kids amounts to a heavy toll but, being a smaller species, predation can take on a more important role. Studies in Sweden have concluded that red fox abundance had a direct connection with kid mortality rates and was indeed the main cause in areas where fox numbers were highest. Similarly, it was noted that at a time when there was a considerable resurgence in roe numbers, this coincided with a reduction of the fox population by a major outbreak of mange.[9]

Wild deer are generally very resistant to disease under normal circumstances. Where populations are allowed to become excessive, however, the build-up of parasites and incidence of disease become more likely: lung worm and liver fluke are particularly associated with conditions of over-population. These internal parasites are themselves strongly associated with parasitic pneumonia, which tends to affect the younger members of deer populations the hardest. Older deer appear to be more resistant.

Accidents cannot be discounted. Apart from collisions with road vehicles, a major killer, deer frequently become trapped (usually by the hind leg) when attempting to jump stock fencing. Grass or silage cutting and harvesting can also be regular causes of death, the times for these activities often coinciding with the presence of young animals whose natural reaction to danger is to freeze and rely on their immobility, rather than flight, as a defence against being detected by predators.

6 Froy, H. et al (2016), *Relative costs of offspring sex and offspring survival in a polygynous mammal*, Biology Letters, The Royal Society Publishing; available from www.rsbl.royalsocietypublishing.org/content/12/9/20160417 (Accessed 28 February 2018)

7 Pemberton, J. and Kruuk, L. (2015), *Red deer research on the Isle of Rum NNR: management implications*, Scottish Natural Heritage and University of Edinburgh; available from http://rumdeer.biology.ed.ac.uk/ (Accessed 28 February 2018)

8 Chapman, D. and Chapman, N. (1997), *Fallow Deer*, Machynlleth, Coch-y-bonddu Books

9 Jarnemo, A. (2004), *Neonatal mortality in roe deer*, Uppsala, Swedish University of Agricultural Sciences

Immigration and emigration

Unless comprehensively fenced or walled, deer are not usually constrained by the artificial boundaries decided by man and may come and go as they wish.

The causes of such movement may be entirely natural. The larger deer species tend to live in single sex herds, and mature males and females may only come together for the period of the rut, living apart for the rest of the year. Many miles can be travelled during this time, and deer often appear on, or crossing, ground where they may not be seen for the rest of the year. This is particularly noticeable among red and fallow deer, and management plans may need to take special care to include all of those who have responsibility for the ground covered if deer numbers are to be tackled realistically, or on the other hand not excessively over-shot.

The smaller deer species tend to be more loyal to their year-round home ranges, though travellers (most often young males) are capable of turning up in new areas as growing populations push them out to seek fresh ground. None of the smaller deer will form herds, although roe will often gather in larger (though only loosely associated) groups during the winter to take advantage of locations with better feeding. Do not be misled into thinking that this automatically indicates that more have immigrated onto your ground.

There is also a degree of natural movement, especially in more northern parts, related to summer and winter conditions. Deer may often seek out higher ground in summer to avoid the torment of biting insects, moving lower only at night. In winter the pattern is reversed, as weather conditions force the deer down towards lower ground. At such times there may also be the added attraction of standing crops such as root vegetables, which offer a food source when natural feeding is scarce.

Deer may also be forced to move by more artificial circumstances. An increase in pheasant shooting activity might render woodland less quiet and attractive, large-scale forestry activities can remove swathes of once available habitat, or changes in farming practices (especially when switching from cereals to livestock or vice versa) could render an area more, or less, attractive to deer. Increasingly, building works and other construction projects are claiming ground used by deer in the past, obliging these animals to emigrate elsewhere.

All of these factors, and more, might encourage deer to move onto or off a management area either temporarily or permanently. Once again, it is up to the deer manager to build up a sound knowledge of local trends and planned activities among farmers, foresters and the like; there is no fixed way or forecasting immigration and emigration beyond the application of sound local knowledge.

⬆ *If there are dominant bucks holding territory in this roe doe's home range, her buck kid will probably be forced to move elsewhere when the following spring comes.*

Special local considerations

Local circumstances vary wildly, and factors may exist in some places which need no consideration elsewhere. Two such are suggested below, though of course there may be others.

Management areas adjacent to, or crossed by, road networks may experience casualties from deer/vehicle collisions. In such places an estimate of projected deer numbers lost on the roads may need to be included in the cull plan. Of course it is

10 Olson, D.D. et al (2014), *Vehicle Collisions Cause Differential Age and Sex-Specific Mortality in Mule Deer*, Advances in Ecology, Vol. 2014, Article ID 971809; Available from: www.hindawi.com/journals/ae/2014/971809/20160417 (Accessed 1 March 2018)

11 Parkes, C. & Thornley, J. (2008), *Deer: Law and Liabilities*, Shrewsbury, Swan Hill

impossible to forecast exactly the sex and age of future casualties, but young deer tend to figure more highly, with yearling roe bucks especially liable to being hit when fleeing from aggressive territorial animals during the spring and early summer. One USA study suggests that although a higher proportion of total casualties was female, with some 40% of those killed being animals of less than two years old, buck casualties were as much as three times higher than their actual proportion among the living population.[10] Whatever the case, it is necessary to at least make reasonable provision of expected RTA losses against experience gained over previous years.

An alternative approach is to take no account of likely RTA casualties during planning, but instead to deduct actual figures from the planned cull totals. Beware, however, that a great many deer hit by cars may leave the scene and expire elsewhere in deep cover, where their carcases might never be found. Also note that RTAs are by no means selective.

Poaching can be another localised issue which affects deer populations, although once again it may be impossible to ascertain the exact numbers lost to illegal activity and some lucky areas never seem to experience the problem at all. It is usually most prevalent in places with good road access and quick getaway routes available. The occasional poached deer may not be a major issue as far as maintaining a cull plan is involved, but large-scale activities can have a huge impact. As many as six or seven deer might be taken in a night's shooting, with the carcases being hurriedly thrown into vans for gralloching elsewhere, thus leaving no real evidence of activity or the number, sexes or age categories of deer involved.[11] In such cases the manager can often make no more than an intelligent guess as to what the losses might be.

The unseen percentage

No matter how comprehensive your census has been, it is very likely that there are some deer that will have escaped counting, especially if you are using a method that relies upon visual sightings. In such cases it is necessary to estimate what percentage this might be.

In more open areas, where you might reasonably expect to see animals fairly easily, this percentage may be quite low, but in any but the most unusual circumstances should not be set at less than 10%. Areas containing a large proportion of woodland or forestry may well demand that the percentage unseen needs to be

The elusive nature of deer, especially when combined with heavy levels of cover, can frequently render an accurate assessment of numbers impossible.

more like 30% or possibly even higher. Although it can be difficult to find the right figure to apply during the earlier years of setting management plans, time and experience soon point to what has proven historically accurate.

The best advice if you are unsure is to over-estimate and work to 'worst case' figures. It is always easier to adjust targets downwards than to have to make a radical increase in effort because you suddenly realise that you have planned against an estimation of far fewer deer than there actually are. The examples given in Chapter 6 of under-counts as far apart as Denmark, Poland and Wiltshire provide ample illustration of how easily this can occur.

The Hoffman Pyramid

It is virtually impossible to avoid mention of the Hoffman Pyramid in any discussion of deer management planning, so a brief explanation of what is involved will be made here. Although the method has its followers for actual cull planning, I do not advocate its use under normal circumstances - although it can provide a useful illustration of population dynamics.

In simple terms, any structured deer population will be roughly pyramid-shaped, with a few, very old animals right at the top and a great many more young ones forming the wide base, as shown in the idealised illustration in Figure 1. The age classes in between form the remaining body of the pyramid.

Of course the situation in real life is not always quite as neat as this. Every year, a new birthing (or 'cohort') replaces the base layer of the pyramid, and the rows that already exist are pushed upwards, their ranks being thinned as time goes on by natural mortality and active human control. Figure 2 shows that the ideal pyramid of age classes is in fact misshapen by those which may outnumber what is desirable, and these extend outside the pyramid to represent the surplus that needs to be culled. In the example shown, a slightly greater effort clearly needs to be made towards culling the female side of the population. Sometimes there is no surplus among certain cohorts, as in the 4, 5, 9, 13, 14 and 16-year-old stags and 13-year-old hinds shown. Next year a new cohort will be born and raise the base of the pyramid once again, pushing the remaining age groups up into to an ever-narrowing proportion.

Assumptions do have to be made, starting with an expectation (not always realised) that equal numbers of male and female calves will be born each year. There is also an implied need to accurately identify specific age-groups, both in the field and in the larder, and for this reason many observers prefer to work in more simplistic terms of kids/calves, and young, middle-aged and old deer. The very peak of the pyramid represents the maximum realistic age limit of the species being managed.

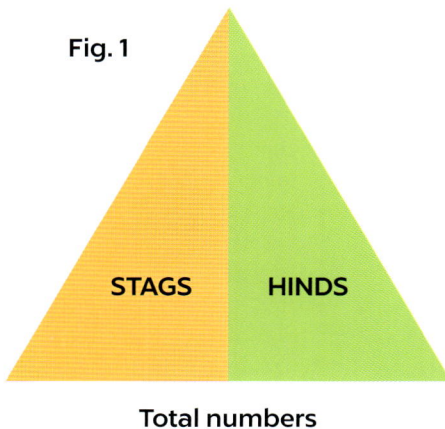

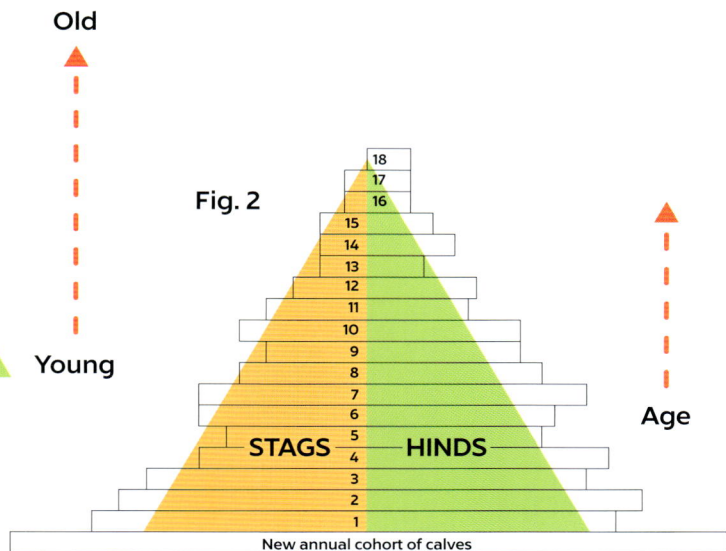

Fig. 1
STAGS   HINDS
Total numbers

Fig. 2
Old
Young
Age
STAGS   HINDS
New annual cohort of calves

There can be very large variations in the shape and structure of the pyramid depending on species, location and other factors. As far as muntjac and fallow are concerned, in many areas where they are shot very few bucks tend to make it beyond the earlier stages of the pyramid because they are often too heavily targeted in comparison with the females. In England and Wales, the fallow buck open season lasts for 9 months, while muntjac of both sexes can be shot all year round. Red deer stags in Scotland, on the other hand, have a very short open season of only around 3½ months and cull animals tend to be taken either as calves or as trophies. Many experienced observers suggest that only a well-managed roe population can really be reflected in a classic Hoffman Pyramid.

Creating a Hoffman Pyramid for your management area and species is an interesting exercise in its own right, and there are several excellent sources of further detail available to those who would like to know more.[12, 13] The description given above is no more than a basic introduction to the process and is included only to give an overview. Many will find it somewhat complicated but happily it is not an essential part of an overall deer management plan. What follows is a much more straightforward approach.

A simple model for cull planning

Having taken a census and considered all of the varying factors that may apply to your management area, it is now time to put them all together and come up with a draft cull plan. The method that follows is based on the one I first learned to use within the Defence Deer Management organisation; it is tried and tested, and more importantly it works.

The exercise will best take place in early spring for a number of reasons, not least that the birthing season with its attendant increase in the overall deer population will shortly be upon you. For those predominantly interested in roe, April also marks the opening of the buck season, which is a good point from which the management year can start.

To illustrate the process, we will use the fictional management area of Cold Comfort Farm, based on 800 hectares (1,978 acres) of mixed farmland with a high proportion of mixed woodland that also supports a small pheasant shoot. The main deer species present is roe, with only the occasional muntjac and a few transient fallow. While the available woodland might support more, a carrying capacity of 60 roe has been agreed to keep numbers at a point where damage to crops is held at an acceptable minimum. Under normal circumstances, an annual reduction of some 30% of the roe population would be necessary to keep numbers stable, but numbers have been allowed to grow beyond this and it is clear that a reduction cull is going to be necessary.

Muntjac are shot as seen, within the ethical guideline of sparing nursing females. Beyond this guideline, it is extremely difficult to create a formal cull plan for muntjac in any case because of the difficulty in assessing their actual numbers to use as a starting point.

The neighbouring estates hold larger numbers of fallow, but it has been agreed with them that only does and young males will be shot according to the legal seasons, and records shared with all interested parties so that they can incorporate them into their cull plans. Mature fallow bucks are seldom seen, and

12 de Nahlik, A.J. (1974), *Deer Management – Improved Herds for Greater Profit*, London, David & Charles

13 Griffith, D. (2011), *Deer Management in the UK*, Shrewsbury, Quiller

in any case a decision has been taken not to shoot these.

There has been a local history of small-scale poaching from an adjacent main road, on which three or four carcases are also found every year after being hit by cars. In addition, the forester on one neighbouring estate has announced his intention to conduct some clearance operations during the coming year.

Now a draft cull figure can be calculated. It is referred to as 'draft' because it should never be considered absolute, and adjustments can be made if circumstances change as the year progresses. Whilst making your calculations, note that you may arrive at fractional figures in some cases. Where these occur, always round the figures up or assume an extra female (rather than a male) if you get an uneven number of kids – in this way you are working to 'worst case' figures which are easily adjusted downwards during practical culling operations if that becomes necessary.

The notes that follow should be read in conjunction with the example draft cull plan on the page overleaf.

Part A

A census taken from observation points was carried out during March, backed up by several night outings with a thermal imager, concluding with a total of 24 roe bucks and 30 roe does seen (Step 1). A further 10% of these numbers have been added to account for deer not seen (Step 2) during the census. This provides a starting point of 27 bucks and 33 does assumed to be physically present on the ground (Step 3).

Part B

Of those 33 does, 60% can be expected to produce young during the coming year so that translates as 20 breeding animals (rounded up from 19.8) (Step 4) producing 32 kids at a fecundity rate of 1.6 (Step 5). There is no way of knowing how these kids will be split into male and female, so an assumption of 50:50 is made, with 16 bucks and 16 does added to the overall population (Step 6).

The forestry operations on the neighbouring ground can be expected to move some deer onto Cold Comfort Farm so another four roe might be gained that way. Again, a 50:50 split is assumed for these immigrants (Step 7).

The winter has been mild and no carcases have been found in the woods so mortality is expected to be minimal, but a further 5 bucks and 5 does have been allotted as losses against road traffic accidents and occasional poaching (Step 8).

Now the total gross forecast population can be arrived at by **adding** together the census figure, forecast births and other additions, then **subtracting** forecast losses (Step 9).

Part C

All that remains now is to insert the carrying capacity that has been decided for the ground in question (Step 10) and subtract it from the forecast gross population. This gives the total cull of males and females that will be necessary (Step 11). If a minus figure is arrived at, of course, the ground is carrying less than it can support, so no cull will be necessary.

The cull can be further fine-tuned by deciding the age categories of animals to be removed (Step 12). Where the population is already well balanced, a cull of 60% young, 20% mature and 20% old animals will normally maintain the *status quo*, but if imbalances exist this is an opportunity to correct them. Getting the balance right

DRAFT CULL PLAN

Location: ...... *Cold Comfort Farm* ...... **Species:** ...... *Roe* ......

PART A. Census as at (date): ...... *26 March 2019* ...... **Method:** ...... *Direct Obs and TI* ......

Step		MALE	FEMALE
1	Numbers seen	24	30
2	Add % not seen __10__% (Guide 10-30%) (Notes 1, 2 & 3)	3	3
3	TOTAL (Steps 1 & 2))	27	33

PART B. Theoretical Population Increase

Step		MALE	FEMALE	KIDS (M&F)
4	Number of females that breed __60__% (Guide 60%) (Step 3 x %)		20	
5	Multiply breeding females (Step 4) by fecundity rate used: __1.6__ (see Note 4)			32
6	Divide result of Step 5 into M/F (Guide 1:1)	16	16	
7	Other forecast GAINS (immigration etc)	2	2	
8	Forecast LOSSES (winter/early mortality, poaching, RTA, emigration etc)	Minus __5__	Minus __5__	
9	Gross population by 1 July (Step 3 + Step 6 & 7 minus Step 8)	40	46	

PART C. Draft Cull Plan

Step		MALE	FEMALE	
10	Carrying capacity	30	30	See Note 5
11	Planned Cull (Subtract Step 10 from Step 9)	10	16	
12	Breakdown of Cull: a. Young __60__ % (Guide 60%) b. Mature __20__ % (Guide 20%) c. Old __20__ % (Guide 20%)	 6 2 2	 (10) (3) (3)	See Note 6

is especially important for the male side of the population, for if there are too many of certain age categories unnatural stresses are likely to build up. This is especially true of intensely territorial species such as the roe. The female age breakdown on the other hand (with figures shown in brackets) is usually not as important, and achieving a 60:20:20 cull, while desirable, is not so essential.

This planning method is simple but has been in use for many years and proven effective over time and a wide variety of circumstances. It has the advantage of being very straightforward and intuitive once the basic principles are understood. A blank copy, along with further explanatory notes, can be found at Appendix III.

There are other planning models available, though they tend to take the same factors into account while approaching them in a slightly different manner. Dominic Griffith offers a very considered approach in *Deer Management in the UK*, while some other comprehensive detail can be found within the Deer Initiative's Best Practice Guide.[14] If you have the inclination, it is always interesting to try an alternative approach and compare the conclusions that you obtain.

Do not be shocked if your planned cull seems rather high against the overall deer population. Even when numbers are at an acceptable level, it would not be unusual for it to be as high as some 35% of the total summer (post-birthing) population for some species with higher fecundity rates. Where the cull is noticeably higher than this, perhaps if there is a significant reduction cull to be achieved, care needs to be taken not to put undue pressure on the deer population. You may have to accept that your plans cannot be achieved in the course of a single year and that you must extend the time period for operations. In such cases the emphasis needs to be placed on reducing breeding females, and the buck or stag cull may need to be given a much lower priority for a year or two.

You may also realise that the numbers are simply too much for a single stalker and decide to seek assistance, perhaps through a series of collaborative cull days which enhance opportunities to achieve the desired reduction in a short period of time. It is important to remember though that increased pressure on the deer, especially through intensive foot-stalking, may persuade them to change their habits and become more difficult to find. Furthermore, inflicting unnecessary stress levels is almost inevitably reflected in overall quality levels within the herd.

It is worth stressing that, with a little understanding, cull planning need not be complicated provided that you follow the proper steps. As long as you know what you are trying to achieve, understand the factors that affect deer population growth and start with a sensible census figure in mind, you are heading in the right direction. Deer management is never going to be an exact science, so what remains is the application of observation, common sense and a bit of flexibility when necessary. ■

14 The Deer Initiative (2009) *Best Practice Guides – Deer Management* available from www.thedeerinitiative.co.uk/best_practice/deer_management.php (Accessed 8 March 2018)

⬇ *Where management is light or does not take place, the larger and more social species of deer can form very large herds. Heavy culling pressure can cause herds to break up and the animals within them to change their habits.*

8 Miscellaneous Matters

The management of deer, and the motives of those who have responsibility for it, are often the subject of unreasonable misunderstandings. If the process is carried out properly, it should be largely unobtrusive as far as those who inhabit, visit or work in an area are concerned. As crepuscular animals, most deer movements will take place either very early in the day or towards last light, when human activities tend to have ceased. Active stalking, being conducted either quietly on foot or from static locations, is by its very nature inconspicuous, so only the shot itself and subsequent carcase extraction have the potential to draw attention to the manager. Even these impacts can be minimised with a little forethought. The widespread availability of sound moderators for sporting firearms reduces noise disturbance and in any case the report of a full-bore rifle being discharged is unlikely to draw attention as little as a few hundred metres away if in woodland.

Other land users

It is easy to slip into a proprietorial way of thinking when managing a particular piece of land, and equally easy to forget that there are others with responsibilities within it. Deer may only be a very small part of the mosaic of land management activities going on and their management may indeed represent only a relatively minor (if important) function within it. Few have the luxury of owning the ground they stalk upon and thus are able to set priorities for it. Furthermore, while the deer manager may take pleasure in seeing deer about, others may not and it is important to give the impression of conducting a serious job conscientiously while taking a humane, professional approach.

In most cases, the deer manager is working within a team which also includes other land users such as farmers, forestry workers and many others, all of whom will come to see him as a positive asset as long as there is honesty, mutual respect and co-operation between all concerned. Most will appreciate having an extra set or two of eyes on the ground, especially outside normal working hours. Reports of suspicious behaviour or vehicles parked in unusual places are always appreciated, as are reports of any loose stock or damaged field boundaries. Strict adherence to the Country Code is essential; take care to leave gates closed and avoid creating damage. If you do, make sure that it is reported as soon as possible with an apology.

↓ The chances are that the deer manager will not be the only person present on the ground, even outside normal working hours.

Other land users will need to be told when you are out on the ground for safety reasons, and of course it will make your job easier if an area is not unwittingly disturbed. It is easy enough to call or text ahead and warn the relevant people of your plans, but very often all that is necessary can be a simple pre-agreed signal that management activities are taking place. Whilst this could involve a small flag or sign in a suitable place, sometimes nothing more than a rock placed on top of a fence post and removed at the end of the outing is all that is needed.

Where there is any degree of public access, more obvious measures may need to be taken when management activities are taking place. While it may not be possible to physically close public footpaths and other rights of way, those using them should be warned, especially if there is likely to be shooting involved. There is no need to be too explicit about what is going on; appropriate notices might be worded along the lines of 'Wildlife management in progress - please do not leave the path', 'Forestry and wildlife management in progress, please keep to the footpath and keep dogs under close control' or simply 'Shooting in progress', depending on the circumstances. A notice might also provide further details of an appropriate person to whom any queries or concerns should be addressed, or simply a contact number. It is often counter-productive to suggest that there is any particular risk to the public, or to make it obvious that animals are being culled, as there will be those who might be unduly alarmed or distressed by the implication.

However you decide to word any signs that you put up, it pays to take care and think through how they might be interpreted. On more than one occasion, arbitrary and blunt notices posted around public access woodland instructing the public to stay out as deer culling was in progress have resulted in unnecessary levels of local outrage and alienation which might have been avoided with a little more tact and diplomacy. At the other end of the scale there is another story, sadly probably apocryphal, of one Highland estate that was troubled by walkers wandering off the marked paths and across delicate moorland. Simple signs stating 'Beware of adders' was all that it took to persuade people to stay on the tracks.

Otherwise, simply be aware of other land users and show respect for what they do. Don't block access routes, don't leave rubbish behind and ensure that all gralloch is removed or buried so that it cannot be found by a passer-by. Firing a shot (especially one which is not moderated) close to residential areas should be avoided as much as possible, and especially very early in the morning or late at night.

Game shooting interests often unfairly view deer stalkers as damaging to their interests. In reality nothing should be further from the truth, as long as roosting birds or game crops are treated with respect and care is taken not to disturb them, especially before shoot days. A helpful and friendly approach will bring rewards in the form of physical assistance when creating shooting lanes or building infrastructure such as high seats, or information on deer sightings and any unusual observations. Local gamekeepers will probably be grateful for foxes to be shot if you are prepared to do so, but always check first as the subject can arouse strong feelings.

Above all, be open and honest. Mistakes happen and will almost inevitably be uncovered quickly if you try to conceal them. It is far better that estate staff are made aware of a potentially wounded animal which has not been recovered rather than to come across it while out on their daily business. Remember, even in the depths of the country, the chances are that you are being watched. Always take pains to work ethically and within the law; if you don't, it will quickly become apparent to those around you.

⬇ Signage should be worded to avoid causing unnecessary alarm or upset.

← Gamekeepers will usually be happy to assist the deer manager, especially where deer impact on their activities. Despite a wire barrier, this roe doe has learned to take grain from a pheasant feeder.

It has been said many times before but the mantra is worth repeating – it can take years to build up a good reputation, but only seconds to destroy it.

Urban considerations

Deer living within urban environments, or on the edges of them, can be the cause of particular problems that need to be approached with special care. While muntjac and roe are the most likely to be encountered, there are increasing instances of the larger species entering built-up areas with the attendant risk of deer/vehicle collisions, disruption, garden damage and other impacts on human activities. Additionally, there are suggestions that deer living within urban areas may pose a health hazard to humans, and there are concerns that established urban populations are often in a poorer physical condition than those living in more natural habitats.

Dealing with deer in places close to human populations, especially large ones, is fraught with hazard. The use of firearms is usually inadvisable, and often impossible, for reasons of safety, disturbance and local sensibilities. In the United States, where the resurgence of white-tailed deer has caused major local problems in some areas, the use of archery as a method of control has met with some success, an arrow being quiet and having a limited fall-out range compared with a bullet. This is not an option in the UK under existing laws, which prohibit the practice for the control of any wildlife.

Deer, as we have considered elsewhere, can create wildly differing emotions among observers, and this issue is magnified in urban areas. The seemingly inevitable outraged reactions to the regular cull of the deer in Richmond Park, carried out under strictly controlled conditions by highly skilled rangers, are a perennial example.

→ *Red deer in Richmond Park. Culling close to any human habitation, let alone densely populated urban areas, demands very careful consideration and is often best avoided.*

Even where it is possible to carry out the shooting of deer in built-up areas – perhaps on the edges of housing areas adjacent to uninhabited land or in villages where a house has no close neighbour – great care must be exercised. Complaints to the authorities will always be investigated closely whenever firearms are involved, and if the user is proven to have acted irresponsibly or outside the conditions of their Firearm Certificate, the loss of the latter and criminal charges may well result. If requested to consider shooting deer close to populated areas, deer managers are strongly advised to seek the advice of their police firearms licensing department to ensure that they are operating fully within the law.

Removal of the animals unharmed may be possible in specific instances, but this can be expensive, time-consuming and highly stressful to the deer themselves. If there are others in the area, you are likely just to be creating a vacuum, which will quickly be filled again anyway. It is probably only practical in situations where deer have become inadvertently fenced in or otherwise trapped, and creating a way for them to depart under their own steam is not an acceptable alternative. Beware, though, that darting with tranquillisers should only be conducted by appropriately licensed and equipped operators, while other means of capture will also require a licence from the relevant authority. Handling deer always demands great skill and care if the animal is not to suffer life-threatening injuries in the process.

It has to be accepted that deer have become firmly established within some urban areas and that for a variety of reasons lethal control may simply not be a realistic option. In such cases the best approach is to fall back on a suitable passive method for exclusion, individual protection measures or deterrents, as considered in Chapter 3.

Dealing with the general public

It is often said that while dealing with deer themselves is relatively easy, it is almost invariably people that will cause the most problems for the deer manager. This is sadly an elemental truth; 'people problems' come in a number of forms and need to be considered differently. Some are simply dealt with but others not so.

The 'right to roam' is a widely misunderstood concept and it must be accepted that the British public, as has often been pointed out, are capable of turning up in the oddest places, at the oddest times, doing the oddest things. As a result, all legitimate land-users must be prepared to deal with the unexpected presence of people who, rightly or wrongly, assume an absolute right to be there.

Even before the advent of the Countryside and Rights of Way (CROW) Act 2000, the country was criss-crossed by a variety of public footpaths, bridleways and other routes permitting public access. The CROW Act, which only applies in England and Wales, implemented the so-called 'right to roam' on specified upland and uncultivated areas named as 'mountain, moor, heath and down' in addition to registered common land. Not all uncultivated land is covered by the Act, however, and this is a common misunderstanding among some who seem to assume more freedoms than it actually allows. Local access forums are empowered to advise on the development of path networks, land access and footpath improvement, involving landowners, users and other interested parties.

CROW is largely echoed by the Land Reform (Scotland) Act 2003, although in Scotland there had already been a long tradition of access to the open countryside and the new Act specifically required that care be taken not to interfere with farming, game or other interests.

The deer manager may well find himself in a difficult position when suddenly confronted by a member of the general public. The presence of a rifle alone may be alarming to someone who is not used to them, and if it is carried by a masked and camouflaged figure such concern is liable to be multiplied. It is as well to consider your appearance carefully and avoid masks or balaclavas altogether, or at least to remove them before making your presence known. Camouflage clothing may have paramilitary connotations, so plain green stalking jackets and trousers are preferable in places where the public are likely to be encountered. Knives are best kept out of sight, and firearms should be carried on the shoulder or, better yet, placed in a slip. If you are in the process of extracting a carcase it should be kept out of public view as much as practically possible.

If it is impossible to avoid an encounter, a polite approach is essential. There are those who object to deer being killed no matter how good the reasons might be; it is best to respect such opinions but be prepared to explain why it is necessary should the chance arise. This is often a good time to stress an approach tempered by humane considerations and a deep respect for the deer, while explaining the

↓ *Recreational use of the countryside is increasing, with potential for conflicts and misunderstandings.*

possible consequences of not keeping numbers under control. Many people are surprisingly open to sensible explanations but some will simply not want to know, so be aware that forcing information on them could be counter-productive. The word 'stalker' carries unpleasant modern implications so the term 'ranger' or similar may be a better way to describe yourself. If possible, carry an official looking identification badge which can be attached to your clothing. These are easily produced using local resources, bearing a digital passport-style photograph, and will help to reinforce your position. A letter of authority, signed by the landowner or their agent, can also be helpful to establish your credentials and explain your presence and activities if necessary.

An over-officious or belligerent approach tends to alienate people, while a firm but polite redirection towards a right of way is usually all that is needed. At all times the stalker needs to be fully aware of the estate's approach to unwanted visitors as well; they can always be referred to a suitable representative of the landowner if they insist on taking matters further.

Where the public may be legally present, do take care not to leave overt signs of culling on the ground. High seats and other infrastructure should be sited well out of sight; this in itself reduces the chance of an unauthorised person climbing them and possibly being injured as well as the potential for vandalism or more malicious deliberate sabotage. If possible, high seat ladders should be removable or planks secured over the rungs to deter access – and of course warning signs are essential. Even a carefully hidden gralloch may be found and retrieved by dogs, though of course it should have been properly buried in the first place. Any preparation of carcases should be conducted well away from public footpaths, not just to avoid being discovered while it is being carried out, but also to avoid leaving quantities of blood and other signs in plain view of a casual observer who may be offended or even assume that something more sinister has taken place.

Where public presence is unavoidable, all is not lost and a positive approach is best. The careful maintenance of public footpaths tends to keep people on them and reduce any tendency to wander off elsewhere, and such tracks can be enhanced by judicious signposting or other waymarks. Many organisations already mark trails with posts and other markers for precisely this purpose, even providing forest activity areas for children, items of interest or information boards, and purpose-made areas for picnics with benches or even barbecues along the way. All of these show an acceptance that there will be a public presence but positively channels it into places where it has the least impact. If physical barriers have to be erected they need to be substantial and constructed of timber or heavy rope. Light tape should be avoided as it can break, be deliberately ripped down, or become entangled with the antlers of deer.

⬇ *Overt signs of stalking activity are best avoided in places where the public may be present.*

← Well-marked nature trails with objects of interest along the way can help to direct and educate the public.

Many larger public access areas even provide parking areas with information boards identifying the local wildlife and activities surrounding it. Although this is good proactive practice in places where the public are encouraged, on more private ground it is probably best kept to an essential minimum or avoided altogether.

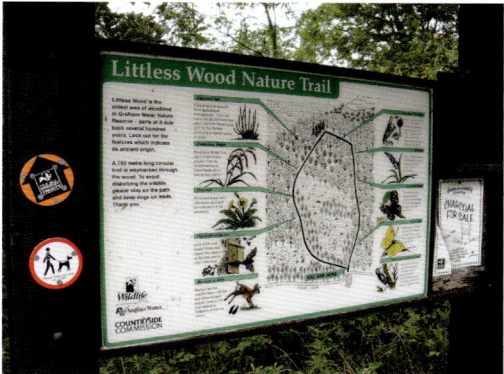

← A tourist information board in a public access woodland car park.

Night shooting

Under normal circumstances, night shooting should be viewed only as a very last resort, even if the necessary licences can be obtained in the first place. Unless the shooting is being specifically done to prevent suffering, such as the humane despatch of an injured animal, authorisation must always be sought. Generally speaking, night shooting licences are not easily obtained. Before any licence is granted, the issuing authority (see Chapter 10) will need to be fully satisfied that there is no risk to public safety, and that there is a suitable and good reason for it to take place. There will be an expectation that the person doing the shooting will have sufficient expertise and experience to carry it out safely.

From the outset it must be emphasised that it is impossible to be truly selective when culling deer using a spotlight. Often there is insufficient time to evaluate individual animals carefully enough before deciding to shoot, and if anything goes wrong a follow-up will be hampered by the darkness. Deer welfare – consideration for which ought always to be at the forefront of operations, as we have seen – may therefore be compromised more easily.

Safety is of paramount importance and a careful risk assessment needs to take place before any attempted shot. Considerable self-discipline is required and if there is any doubt about an effective backstop, or about what else might be in the vicinity of the target, it would be foolhardy in the extreme to release a bullet. Ricochets are a special hazard when it is difficult to identify fences, rocks or other hard objects that might deflect a shot. It follows that intimate knowledge of the land is required, and it must always be borne in mind that there may still be estate workers or members of the public in the area, even after dark. There have been incidents in the past where birdwatchers and the like have been mistaken for animals because of the light reflecting off tinted binocular lenses. Working at night tends to magnify the potential for mistakes to be made, so working practices need to be carefully agreed by all involved beforehand, and first-aid kits and appropriate communications need to be available throughout.

It is advisable to warn anyone living or working in the general area that night shooting is taking place, and what the start and finish times will be. While usually not legally required, it is also sensible to tell the police beforehand if a visit from an armed response unit is not to be triggered by a call from a concerned citizen. Further guidance on best practice and the conduct of night shooting is available from the Deer Initiative[1] or Scottish Natural Heritage.[2]

Unless it is conducted with extreme care, night shooting is fraught with potential hazards and it is strongly recommended that it should be avoided if at all possible. If results are not being achieved during normal daylight hours, it may be necessary to conduct an honest review, not just of what you are doing but also of the resources you have available.

Getting help

Too many people seem to take on more deer ground than one person can reasonably cope with. Most deer managers unfortunately have to work elsewhere to make a living, although a fortunate few are able to do the job full-time. It is very easy to over-commit yourself and lose sight of the fact that this is a serious enterprise. Once a cull plan has been agreed it needs to be fulfilled, and it is unlikely that a landowner will be impressed if for whatever reason – be it accident, illness or something as simple as a family holiday – agreed targets are not reached.

Sometimes it is simply not possible for a single-handed stalker to cover large amounts of ground or meet large cull targets alone and reality demands that assistance be sought. While a full-time professional might reasonably expect to conduct an annual cull involving several hundred animals, a part-time manager with only restricted times when he can be on the ground might struggle to achieve two or three dozen. There are always people looking for stalking and, as long as care is taken to choose trustworthy individuals who can be relied upon, a bit of help will make the manager's job much easier. Such assistants need to be

1 The Deer Initiative (2009) *Best Practice Guide – Night Shooting*; available from www.thedeerinitiative.co.uk/uploads/guides/92 (Accessed 25 April 2018)

2 Scottish Natural Heritage, *Best Practice on the Management of Wild Deer in Scotland – Night shooting*; available from www.bestpracticeguides.org.uk/reference/night-shooting (Accessed 25 April 2018)

← *Faced with a large cull, or large areas to cover, it may be necessary to seek assistance.*

dependable as well as competent, and while holding a suitable qualification such as DSC1 or 2 is strongly advised it is even better if individuals are already known to the manager or a mutual friend who can vouch for them. There will of course need to be an initial degree of supervision but trust should build over time.

Before an assistant is taken on, a few ground rules need to be established. At the very least, access to the ground needs to be considered, and it must be made plain that the person has to work within the agreed cull plan. Indeed, the understanding could be that help is only needed for culling does and hinds during the winter months, with perhaps the occasional summer outing as a 'thank you'. It is also a sensible precaution to ask for evidence that any assistant's firearms are covered by a Firearm Certificate (and that the conditions on it permit it to be used on the ground in question and for the intended quarry species), and that appropriate shooting insurance is also held.

If there proves to be any evidence of bad practice or illegal activity there must be a strict understanding that all arrangements would cease immediately. Should the partnership not work out, it helps if it is possible to dispense with an assistant's services simply and without any unnecessary legal fall-out or bad feeling. This is easiest if no money changes hands; and it does not matter if the assistant is a recreational stalker with no interest in management – what counts is that they can be relied upon to operate safely, achieve the desired results and cull only what they are meant to. Most people would be happy with such an arrangement, as it provides them with free stalking and perhaps the occasional carcase to take home in return for their commitment and time.

Stalking leases

The value of stalking and the venison produced as a result has already been considered in Chapter 4, and a few caveats associated with selling stalking rights raised. If a landowner is determined on this course it is wise to implement a formal stalking lease. In the UK it is normal for such an arrangement to run for not less than five years, with a provision for annual review and a requirement for a fair amount of notice should either party, be it the landowner or the leaser, wish to end it.

A lease will formalise what is permitted and will define what the responsibilities of each party are within its terms. There are many aspects that it might consider, not least the ownership of any venison or trophies that are produced as a result of management, and how any proceeds are to be divided. Importantly, it will need to cover whether the lessee (the person to whom the lease has been granted) is permitted assistants or paying guests, or whether they may even sub-let. To ensure that the landlord has met their own duty of care, a specification of minimum training or qualification requirements may be desirable and if there is liability attached to any damage incurred as a result of management activity. Additional considerations may also be found within the simple letter of permission example provided at Appendix VI, although they may need to be expanded upon.

A formal stalking lease provides peace of mind to all involved and enables the landowner to retain an acceptable degree of control. It also protects the lessee; all too often you hear of deer managers, working within an informal agreement, losing their stalking rights after long periods of time and the efforts associated with implementing a constructive long-term plan, after the landowner has received a financially lucrative offer from someone else. A sensibly thought-out lease is a legally binding contract which helps to ensure continuity of management within an agreed framework and works to the benefit of everyone, as well as that of the deer themselves.

Stalking guests

It may also be tempting to try to derive an income from irregular paying guests. From the landowner's point of view this may initially be an attractive option, but it

➜ *A misty Hampshire morning in late July. In southern parts of Britain at least, many consider the roe rut to represent peak season for roebuck stalking.*

will in itself incur costs: an appropriate person will need to be employed on a full- or part-time basis to guide any guests and ensure that their activities are safe. Sometimes a gamekeeper or other estate worker may be in a position to take on such a responsibility, but more often than not it would amount to a considerable imposition in addition to their normal duties. Roe buck stalking in particular clashes heavily with the most important period of the game rearing season, while other peak times may coincide with game shooting itself. Guiding clients demands special skills and cannot be entrusted to someone who does not have considerable experience of stalking and shooting deer themselves.

A morning or evening stalk in a lowland situation will probably cost a paying guest in the region of £80 to £100 (at 2019 prices), while a day on the hill attempting to stalk and shoot a stag might cost some £400 a day or more. Actual costs may vary considerably from place to place according to local circumstances, and there may be additional charges for the trophy itself as well as other associated costs. Stalking hinds and does as part of the winter cull tends to be somewhat less expensive. Guests will expect an acceptable return for their payment. As well as the actual outing, they will certainly expect their guide to be hospitable, knowledgeable and competent enough to put them in front of a suitable animal.

A good hit on the 'iron stag' on a Scottish sporting estate. Shooting guests need facilities to check their rifles before shooting at a live deer.

If this seems an attractive way of recouping costs or even producing an income, there are drawbacks that must be factored in. Stalking guests are often unknown quantities and may vary considerably in terms of ability, expectations, experience and even fitness. Their demands may vary widely between specific trophy requirements, a simple introductory experience, training or a witnessed stalk for their DSC2 portfolio. For the most part, whoever it is will need to be accompanied at all times. Success rates for outings will not usually be as high as those undertaken by a seasoned, regular deer manager who is fully familiar with the ground and the deer it contains.

To accommodate paying guests, a more complete and formal infrastructure will be required compared to the one required for regular deer management operations. Quite apart from equipment and transport considerations, serious thought will have to be given to adequate insurance, whilst access to a good tracking dog will be essential in case of misplaced shots. A suitable rifle range area will need to be identified; guests will not only need to check the zero of their rifles before outings but their host will need to be reassured of their shooting abilities. To avoid later disagreements, a carefully considered contract must be agreed between all parties, considering not just outing costs but also carcase fees, trophy charges, lost animals and a host of other eventualities.

Finally, do not underestimate the actual time involved where escorting guests is concerned. Quite apart from normal day-to-day transport, the professional will find themselves working some very long hours to accommodate paying guests and ensure that they receive the experience that they have paid for. During the summer especially, an extended run of days involving being up long before dawn, preparing trophies and completing other essential tasks, and getting to bed some hours after dusk, can leave the professional exhausted in no time.

A simple costing exercise will quickly highlight the fact that financial margins may be tight for a landowner or professional deer manager relying on an income from sporting guests. It is not an undertaking to enter into lightly and a discussion with an established professional will quickly place the realities of such a venture into perspective.

Deer management groups

Deer are no respecters of man-made boundaries and can range across large areas, taking in any number of properties as they do so. The larger herding species are more prone to this, while the smaller roe, muntjac and Chinese water deer tend to be more loyal to their rather more restricted home ranges. Efforts to achieve an effective and balanced cull and maintain a structured population can be extremely frustrating when neighbouring landowners are working towards different objectives, especially if they have none at all in respect to the resident deer.

A personal example was given in Chapter 7, where a local mixture of indiscriminate shooting, trophy hunting and no control at all made it impossible to set and achieve a desired management plan. Similar situations occur all too frequently, especially when one estate aspires to maximising its deer numbers to achieve, for instance, a high sporting income, whilst next door a major natural regeneration project is being thwarted by a failure to keep deer numbers down to an acceptable level. The almost inevitable result of such a conflict of interests is recrimination and bad feeling.

It follows that co-operation between neighbours is essential for effective deer management to be achieved. Sometimes informal contact may be enough, but when multiple interests are involved it may be more helpful to form a Deer Management Group (DMG) which provides a forum to agree objectives, while at the same time sharing and promoting best practice, encouraging communication and generally fostering a more co-operative approach. A DMG might be set up for a very specific objective, such as managing an existing problem or even averting one before it arises. Alternatively, its aims might be much more wide-ranging and flexible. When aims conflict, face-to-face meetings often make it easier to achieve a compromise than a reliance on letters, emails and infrequent telephone calls. Many DMGs exist already, but they tend not to advertise their presence too widely. To find out about any in your area, local enquiries may bear fruit, otherwise approaching national organisations such as the British Deer Society, Deer Initiative, Forestry Commission or Scottish National Heritage (see Appendix II) will usually yield suitable contacts.

If no DMG exists already, setting one up is in itself relatively straightforward – but may require a great deal of diplomacy! For the group to work, it will need the involvement of all who have an interest in deer in the area. Suggested interested parties might include landowners or their managers, stalkers, game shooters, farmers and wildlife groups. It can be especially useful to involve a police Wildlife Crime Officer if one is available. The key is to involve as many decision-makers as possible, whilst keeping group membership to a sensible level. A suitable person should be invited to act as Chair: whoever is chosen need not necessarily be actively involved in deer management but does need to have a good working knowledge, possess good communication skills and – most importantly – must enjoy the trust and respect of all group members.

How often a DMG meets depends on specific local needs but should not be less than once a year. Ideally this will be before annual cull planning commences, so that individual members can go away with a good idea of what has been collectively agreed, and accordingly what they are able to achieve in their own areas of responsibility. Meetings can range from a simple, informal gathering in the local pub to a more formal affair. If necessary, specialist advisers can be invited to attend as required. Often these meetings become a keenly anticipated social occasion and can provide an opportunity to invite outside speakers, thus adding to everyone's knowledge and interest. Links between neighbours frequently extend to cross-border assistance with management activities or shared facilities, enhancing effectiveness while reducing overall costs.

In addition to a Chair, the DMG should have a secretary who can take notes and circulate records of what has been agreed to all the members. Care should be taken to ensure that business is kept strictly to the point: it might include large-scale management planning, collating cull data, assessing deer movements and trends or promoting best practice. Discussions need to be controlled, as too much irrelevant divergence from the true business in hand tends to generate impatience and dissatisfaction amongst those present. The Deer Initiative is a good source of detailed guidance on the conduct of DMGs.[3]

3 The Deer Initiative (2009) *Best Practice Guide – Deer Management Groups*; available from www.thedeerinitiative.co.uk/uploads/guides/116. (Accessed 23 April 2018)

Good communication between group members is vital and maintains both interest and unity of purpose. As well as actual meetings, newsletters are a useful way of ensuring that information is passed on; as most people have access to email, costs can be kept to a minimum. A degree of confidentiality is essential, however, as business information can be sensitive. Of course data protection legislation will have a strict bearing on the storage and use of personal details too.

Co-operative/group culls

While most deer management operations will be single-handed efforts, there are occasions when teamwork can pay dividends (as we saw in Chapter 6, when conducting a census). Well-planned co-operative culls can achieve good results when circumstances permit them. Large culls, which might otherwise demand disproportionate levels of time and manpower, can be completed more quickly in such a way. Indeed, for this reason many managers prefer them: the fact that the shooting is completed over a shorter period of time results in far less stress on the deer population itself. It is an interesting fact that managers of park deer often deliberately use unmoderated rifles for culling, as the deer seem to know instinctively that a danger period is over once the shooting has clearly stopped. Many also comment that a side benefit of unmoderated shooting is that the public are quickly made aware that it is taking place and accordingly avoid the area.

A co-operative cull can involve no more than a handful of rifles, and even just two people covering a limited area of land can achieve considerable success, but the best results are more often obtained where larger numbers are employed. Ideally, and within the bounds of safety, as much of the ground as possible should be covered either from high seats or suitable vantage points.

It may be decided to move the deer, which permits the operation to take place at times of the day when they are usually inactive or in places where they might be hard to access or see. The deer themselves are not driven, as they might be in a formal continental-style hunt, but instead are encouraged to move quietly along their regular paths by one or two walkers (wearing high visibility jackets for safety purposes), who make minimal noise but instead allow their scent to drift into bedding areas and persuade the deer to vacate them. In some exceptional circumstances, steady and reliable dogs can be used to push out areas of dense cover. The desired result is for the deer to move away slowly rather than fleeing blindly, stopping frequently to look around them; this permits those charged with the shooting to assess more carefully the cull animals before taking a steady, aimed shot. Under normal circumstances only static animals should be taken, as firing at moving ones is more likely to result in unnecessary wounding.

It follows that anyone with a rifle on such days must be entirely competent and can be relied upon to act safely and within the bounds of what is asked of them. At the end of a move, they should be able to indicate the location of all shot animals as well as give an informed opinion on any that might have made off. Having trained and proven tracking dogs on hand is essential, as it is entirely possible that an animal that has not fallen immediately will need to be followed up and recovered.

If there is public access to the ground, special attention needs to be given to displaying appropriate notices and, if necessary, posting lookouts or assistants at access points to advise of any unexpected public presence. Where possible, the lookouts should prevent access or at least diplomatically explain what is happening (advice on dealing with hostile members of the public is offered further on in this chapter).

Planning will need to be meticulous if all is to go well. The day should start with a comprehensive briefing for everyone involved; suggested notes on what should be included are to be found at Appendix VII (these notes will also assist with pre-planning). Special care needs to be taken to make it clear as to who has particular responsibilities, such as those in charge of various aspects of the day, as well as details of what to do in any emergency. You also need to run through the risk assessment with everyone, which will itself have been completed well beforehand.

Wherever firearms are involved, safety instructions must be stressed as being of paramount importance – and this part of the briefing goes hand-in-hand with the identification of the siting of shooting positions and permissible arcs of fire.

Good organisation is obviously essential and disciplined radio communication is of great assistance in this. Carcases are usually left where they have fallen until the end of the move. As there are likely to be a number of these, attention should be given beforehand to their efficient recovery and processing; appropriate transport needs to be on hand and, if necessary, game dealers liaised with in advance.

It goes without saying that appropriate insurance is a prerequisite, and a physical check of the Firearm Certificates of all those using rifles is strongly recommended. Finding the right people to assist in all the roles is key and may cause the most headaches: but very often co-operation between neighbouring estates is not only possible but mutually beneficial, as well as being likely to yield the best results for everyone.

↓ *The aim of a 'moved day' is to persuade deer to proceed quietly and unhurriedly along their accustomed paths.*

Poaching

Forget the romantic image of a poacher taking a deer as 'one for the pot' to feed a starving family. Modern poachers often operate in well organised and equipped gangs, possessing good local knowledge and even employing diversionary tactics to distract the authorities from the area that they are targeting. While the motivation of some may be little more than the thrill of the chase, there are large profits to be made and ready markets for those who may even treat the activity as a sort of profession. Methods vary across the country but may include anything from the use of snares and running dogs to rifles and lamps or even crossbows. In addition to modern 4x4 vehicles, night vision devices have recently been added to the poacher's armoury. Little consideration is given to public safety.

In all cases, the welfare of the deer is never a priority. It matters not that an animal may be out of season, has a dependent fawn, or indeed whether it is killed quickly or not. Wounded animals are not followed up for humane despatch and damage to tracks, fences or other infrastructure is not seen as important. Carcases are thrown hurriedly into vehicles without being gralloched and little, if any, attention is paid to meat hygiene. Many offences, other than the poaching itself, are associated with the activity.

Despite the seeming indifference of much of the public, poaching is a serious rural crime which impacts on a great many people. In addition, an intelligent, long-term deer management plan can be completely ruined by a spell of indiscriminate, illegal and often large-scale killing.

⬇ *Poaching can range from the opportunistic to the systematic killing of deer by organised gangs.*

Poaching inevitably tends to be more prevalent where there is good road access, which permits the perpetrators to escape the scene more easily. Preventative measures can include barriers and ditches, but often it can be difficult to prevent illegal access to land. Most police forces have dedicated Wildlife Crime Officers and, along with a number of rural organisations, offer advice, seminars or advisory visits to those who are particularly affected by it.

Where incidents occur, the police are far better placed to make arrests and this is a much better option than risking your personal safety, perhaps even chancing counter-charges as well. Incidents should be reported immediately, with as much information as possible passed on to the call-taker. If you are being threatened, or damage is taking place, it is quite justifiable to use the emergency 999 number; in other cases, the non-emergency 101 may be more appropriate. Details given should include as accurate a location as possible (especially at night), numbers of vehicles and people present, and whether firearms are involved. If the police are able to attend immediately, it is often easier to agree an easily located meeting-point from where they can be guided in. Sometimes police attendance may not be immediately forthcoming, but at the very least you should ask for an incident number and ask that the incident is recorded as a rural crime. Poaching tends to be badly under-reported.

If the police are unable to attend, it is still possible to collect evidence to pass on to them. Such evidence might include photographs, videos, vehicle descriptions and registration or scientific evidence in the form of DNA from bone, blood or hair. Keep written or voice-records rather than relying on your memory; if successful prosecutions are to result these are far more reliable. Remember, too, that the best evidence is that which records an unbroken chain of events.

Animal activists

There are, regrettably, some who choose not to operate within the law to pursue their own personal agenda and will do their utmost to disrupt legitimate wildlife management. Such activism is more usually associated with hunting and game shooting, and for the deer manager this is unlikely to involve direct confrontation, but when working alone it pays to be careful. High seats in particular are at risk, and can be either torn down or deliberately sabotaged. Wooden ladders have been known to be specifically targeted, their rungs sawn through on the underside in the hope that they will collapse when someone climbs them. For such reasons it is always wise to site seats well away from public view, and to inspect them carefully before use.

In the unlikely event that you are confronted by an aggressive member of the public while conducting your legitimate management activities, the best advice is to remain calm and measured, and refuse to become involved in an argument. If a firearm is being carried it should be unloaded and placed in a slip, or better yet secured in a vehicle, as soon as possible. Remain aware at all times that you could be being filmed: so be very careful not to react to provocation. This of course can work both ways, as modern mobile phones usually contain digital recording applications which can be used to make a record of actual events in case of a future dispute.

Take care to park your vehicle unobtrusively when working in the field and avoid displaying car stickers which show membership or support of shooting associations. These may cause it to be targeted and vandalised. Most importantly of

➡ *The control of deer can sometimes provoke strong emotions.*

all, it pays not to advertise your activities too widely. In this era of social media, it can be alarmingly simple for others to find and target an individual with whom they disagree, and your own photographs (especially those showing recognisable faces, vehicles and their registration numbers, or easily recognisable locations) are best kept private and not published on Facebook or other public sites.

It can be both intimidating and yes, frightening, to be targeted; if this occurs the police should be informed immediately. Sadly trespass, the act of being on private land without permission, is only a civil offence and the police have no powers to intervene. When circumstances translate into actual criminal damage or assault (which may itself consist of no more than a threat of violence) the police can act if they believe that a criminal offence has been or is about to be committed.

Humane animal despatch

As traffic volumes increase against a burgeoning deer population, the issue of humane animal despatch (HAD) has come very much to the fore in recent years. Euthanising a critically injured wild animal may not be pleasant but can be entirely necessary as it is frequently impractical, not to say downright cruel, to attempt to rescue and rehabilitate it. Wild deer are naturally terrified by the close proximity and scent of humans; one that does not react strongly to either is probably seriously injured, even though it might show few external signs of this. Removing it from the scene may actually contribute to the stress involved, and rapid despatch may indeed be a kinder option.

HAD can be a difficult task to conduct swiftly and competently, and should only be undertaken by skilled and experienced persons. Although many vets and animal welfare charity employees are capable of carrying it out, they tend to be few and far between and a response from them cannot be relied upon. An increasing number of police constabularies have recognised the need for timely HAD call-outs and, rather than relying on their own trained resources, have started to run highly successful schemes whereby invited volunteers can be called out to attend incidents and take quick and appropriate action on their behalf. Call-out lists maintained in operation rooms ensure that suitably located persons can be contacted quickly and directed to where an animal is lying injured. Although the details of schemes can vary between constabularies, in the majority of cases volunteers tend to be provided with high visibility clothing, insurance and, most importantly, training.

A proper form of insurance cover is essential. Those holding personal insurance for countryside activities should still check with their policy provider to ensure that they are covered for such a job, as work-related or commercial activities may not be included in the policy. As noted above, anyone involved in police or other approved schemes may be provided with cover, but the onus is always on the individual to ensure that they are protected in the event of a claim of negligence.

When attending any HAD scenario, the first thing to consider must be the safety of yourself and those around you, even though there may be pressure to get on with the job. Every incident is liable to produce different challenges, and many tend to be on or close to busy roads so traffic can be a major hazard. Add to this the fact that there are likely to be a number of onlookers and the task gets even more difficult. Under ideal circumstances, there may be a police presence to control vehicle flow and keep spectators back but this is not always the case. The HAD operator may well find themselves working alone with no support in the face of difficult conditions and the very real possibility of a hostile public reaction.

A briefing session for HAD call-out volunteers.

A word of caution is in order regarding the handling of injured animals. Even a deer that seems superficially incapacitated may summon enough strength to react vigorously when a human comes into close proximity, and antlers and lashing hooves can cause significant injuries. Muntjac and Chinese water deer may also slash dangerously with their canine tusks, and even the careless handling of dead animals can result in serious cuts if care is not taken with them.

Where a deer is so badly injured that it must be euthanised, the Deer Act no longer applies with regard to permissible calibres, close seasons, hour of the day or other matters. What is important is that the animal is put out of its misery quickly, effectively and in a humane manner. Indeed, a .22 calibre rifle or very small-bore shotgun (such as a .410), both of which would be illegal for the killing of healthy wild deer under normal circumstances, may prove to be ideal for the purpose. Great care must be taken, however, if contemplating the use of any firearm in the proximity of any public place. In many cases a knife or captive bolt device may be more appropriate, but once again it is stressed that in all cases the user needs to be skilled in the use of them. There is no need to obtain specific permission from the authorities to euthanise a wild deer but you do need to remain aware that the law regarding trespass, public safety, animal welfare and other relevant considerations will almost certainly apply. No matter how good your intentions are you will be liable to prosecution if you transgress any of them.

↓ The personal safety of the HAD operative must always be at the forefront of planning. High visibility jackets are an important precaution especially when working near roads, while ear and eye protection (against potential splashback from the shot) should all be regarded as essential.

Great care needs to be taken with regard to public perceptions of HAD and the possibility of adverse reactions. Many people find it very difficult to accept that there is no option other than to destroy a stricken animal, and reactions might range from distress to a forceful intervention. If there is no authority figure, such as a police officer, on hand to control or direct any members of the public, it is possible for a situation to get out of hand very quickly. The HAD operative needs always to be heedful of the emotions that can be aroused, and of course how the act of despatching an animal may appear to an onlooker who is not familiar with the practicalities.

This whole subject is potentially very complex and anyone anticipating becoming involved in HAD would be well advised to find a suitable mentor or, better yet, attend a short course. Various organisations offer these but the British Deer Society's is especially recommended. This was developed primarily for the police and their nominated call-out staff, to demonstrate that all of their operatives, from control room staff to the person actually carrying out the task, have undergone formal training in the roadside despatch of injured deer. It has the credibility of being approved and certificated by Lantra Awards and focuses on best practice, risk assessment and safe operation as well as other relevant issues such as equipment. As you might expect, a major element of the course considers the extremely important legal implications.

Elsewhere, the Deer Initiative and Scottish Natural Heritage provide comprehensive Best Practice Guides which look at procedures, equipment and other aspects in further detail.[4, 5] ■

4 The Deer Initiative (2010) *Best Practice Guide – Deer & Vehicle Collisions*; available from http://www.thedeerinitiative.co.u</uploads/gu des/127.pdf (Accessed 20 August 2018)

5 Scottish Natural Heritage, *Best Practice Guide – Humane Dispatch*; available from h:tps://www.bestpracticeguides.org.uk/culling/humane-dispatch/ (Accessed 201 August 2018)

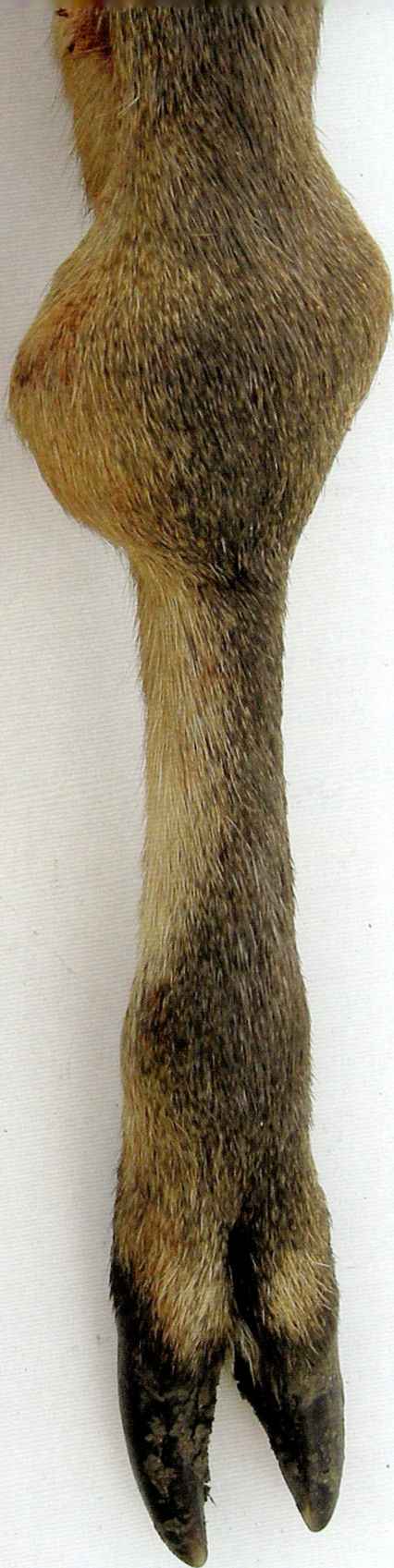

9 Health and Disease

Deer are generally very healthy creatures and, apart from perfectly normal burdens of parasites, it is extremely rare to find one suffering from anything untoward. However, it is still important to know what to look for and what action to take if there are signs of something unusual.

Any living animal displaying evidence of abnormality bears closer watching, and if necessary an expert opinion should be sought. Those familiar with normal deer behaviour should be able to spot anything unusual very quickly and to discount seasonal activities which may have puzzled the layman.

Dead animals should always be inspected carefully, if practicable (it may not be if the animal has been dead for a while, or is lying somewhere inaccessible), and particularly if they have been shot and are about to enter the human food chain. In the latter case the carcase has to be examined by a trained person (see Chapter 4, Active Management) who is appropriately certified as fit to do so.

The stalker should always start with an external examination before commencing the gralloch, looking for anything out of the ordinary – growths, sores, old wounds and the like. Look inside the mouth, paying special attention to the tongue, to see if there are any unusual sores or other marks. Tooth wear will give an idea of the animal's age. It is not unusual to find external parasites in small numbers. Keds move fast all over the body, but lice and ticks, usually found in the warmer areas such as the groin, are more easily spotted. Heavy infestations of lice in deer may be an indication that the animal is sick or in otherwise poor condition or perhaps that there may be a problem with over-population. Tick numbers can tend to rise and fall with the seasons and according to varying land use.

↑ *The appearance and behaviour of a living deer may raise suspicions of poor health.*

As the gralloch progresses, check the animal's internal organs for the unusual, and do not be afraid to seek advice if you are not sure. Treat any scars or discolouration on the internal organs with suspicion, and look out for growth or nodules that seem out of place. That carcase is going into the human food chain, be it for personal consumption or through a game dealer, and responsibility for ensuring that it is fit for the purpose lies with the stalker in the first instance. If in any doubt, samples should be bagged carefully and looked at by a veterinary examiner or other knowledgeable person; if there is any suspicion that a notifiable disease might be present (see section below), then the matter must be reported through the proper channels.

A visual guide to inspecting shot deer carcases is included in a comprehensive DVD available from the British Deer Society. It is highly recommended to anyone with responsibility for this task.[1]

This has been a brief overview of what is expected as Best Practice. What follows is a more detailed look at the diseases and parasites which may affect deer, but first it is important to understand those important indicators of infection, the lymph nodes.

The lymph nodes

The lymphatic system is an enormously important indicator of the general health of any animal, deer included. Tissue fluid bathes all the cells in the body, and makes its way into the bloodstream by one of two routes. One is through the small capillary blood vessels; the other is via the lymph vessel network. Before this fluid enters the blood vessels it has to pass through lymphatic tissues, some of which are formed into small capsules known as nodes. These nodes act as filters for dust, bacteria and any other foreign bodies to stop them entering the bloodstream. Any excessive swelling of the node is an indicator that something might be amiss.

There are seven main sets of lymph nodes, as shown in the accompanying illustrations, that are the concern of anyone inspecting a deer carcase. It is accepted best practice that all should be inspected by the stalker before any carcase passes into the human food chain, as they are important indicators of possible infection.

Lymph node locations

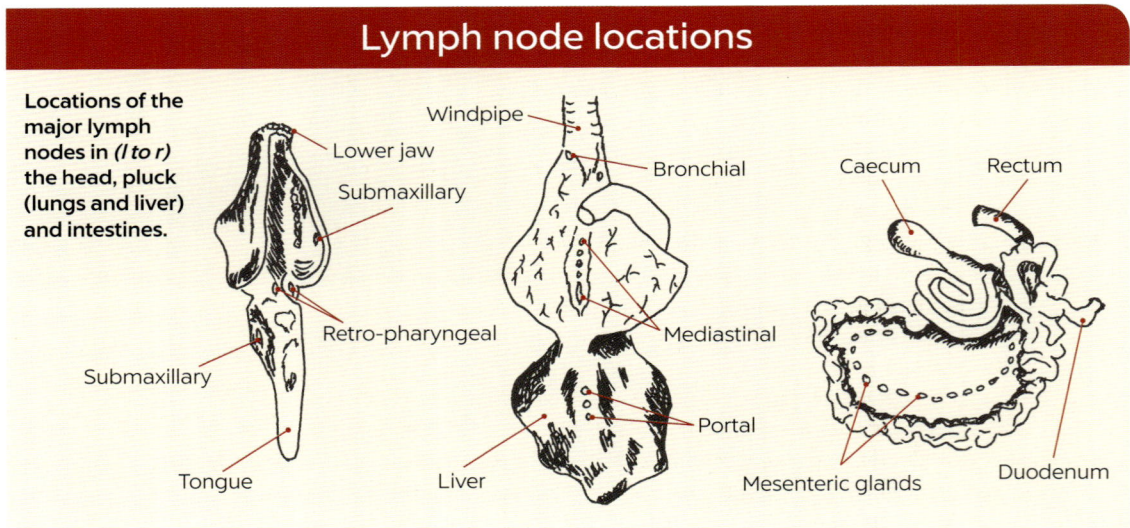

Locations of the major lymph nodes in *(l to r)* the head, pluck (lungs and liver) and intestines.

Lower jaw · Submaxillary · Retro-pharyngeal · Submaxillary · Tongue · Windpipe · Bronchial · Mediastinal · Portal · Liver · Caecum · Rectum · Mesenteric glands · Duodenum

Some of the nodes can be very difficult to find, and it helps to have them pointed out by an experienced person. The easiest to find are the *mesenteric nodes*, which lie in a line between the coils of the intestines. Even if you have no intention of the carcase going any further than your own kitchen, you should get into the habit of checking these nodes as a bare minimum precaution. A classic sign of significant infection is swelling of the mesenterics, especially in bovine TB; the submaxillary and retropharyngeal nodes of the head should also be checked if possible.

1 British Deer Society DVD, *Gralloch: from field to larder*

172

All of the lymph nodes should be smooth, firm and ovoid or bean-shaped. They vary in colour from white to light brown or grey. Any that are obviously swollen, externally mottled or grossly misshapen are a potential cause for concern and demand further examination. They should **not** be cut open to see what the inside looks like: if there is any infection there, doing this will only give it the chance to spread further.

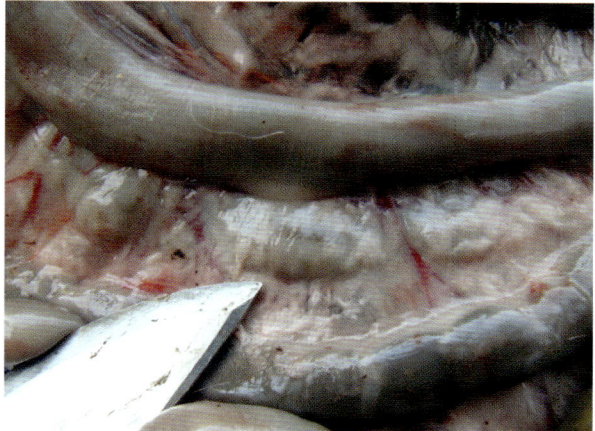

Notifiable diseases

Certain diseases carried by animals must be notified to the authorities so that they can be controlled and, where possible, eradicated. This is a legal requirement under the Animal Health Act 1981 which states that 'any person having in their possession or under their charge an animal affected or suspected of having one of these diseases must, with all practicable speed, notify that fact to a police constable'. Rather than doing that, it has since become best practice for the incidence to be notified to the Animal and Plant Health Agency (APHA).

Notifiable diseases may be classified as exotic (not usually found in the UK), or endemic (normally present here). Up-to-date lists can be found on the APHA website which also notes when a specific disease was last confirmed in this country. It also usefully lists which diseases are considered exotic and which are endemic, and, importantly, whether they are zoonotic (i.e. can pass between animals and humans).[2]

As far as deer are concerned, the main notifiable diseases are anthrax, bovine TB, bluetongue and foot-and-mouth, but there are others and it is important that anyone who has an involvement with deer learns to recognise the signs. Interestingly, warble fly, technically a parasite and not a disease, is also named in the list but the relevant legislation applies only to cattle in Scotland. Many of those diseases identified have never actually occurred in the UK and others have not in living memory. Rinderpest, for example, was last recorded here in 1877 and was believed to have been eradicated worldwide by 2011, but it is still listed. Chronic wasting disease, a major issue in the USA, has never been identified in this country but is becoming of increasing concern since cases, albeit isolated, have appeared in Scandinavia during the past few years.

Anyone who finds a dead deer or examines a shot one which they suspect may be infected with any notifiable disease should contact the local APHA office. What to do next is covered on page 181.

Major diseases of deer

As already noted, wild deer are usually healthy. The smaller species tend not to congregate together and so reduce their susceptibility to infection still further, unlike the larger herding deer and particularly domestic stock. A few, though not all, of the diseases which can affect deer are discussed in more detail overleaf.

↑ *A healthy mesenteric lymph node (indicated by the tip of the knife).*

2 Animal Plant and Health Agency, Notifiable Diseases https://www.gov.uk/government/collections/notifiable-diseases-in-animals#list-of-notifiable-diseases

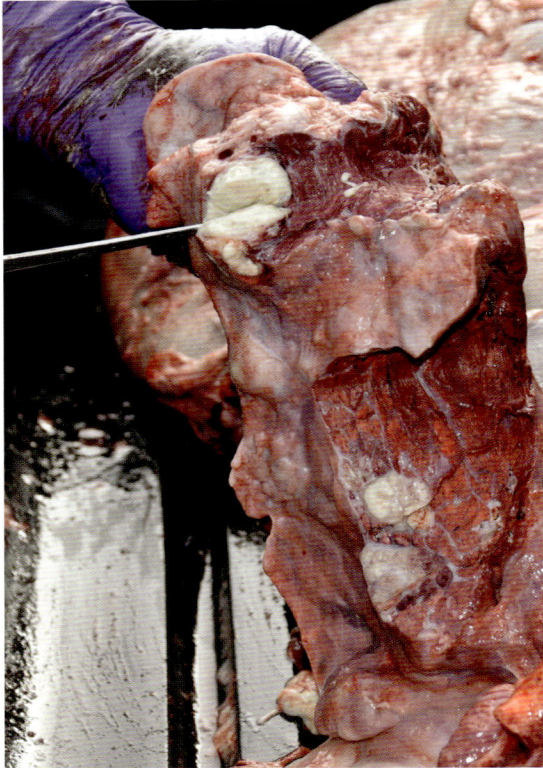

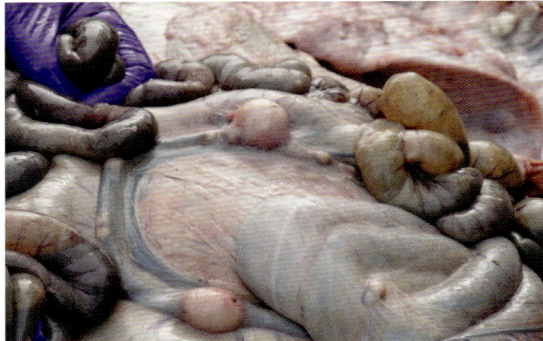

Deer can certainly suffer from **bovine tuberculosis (bTB)**, a serious disease of animals and humans caused when the TB bacteria form multiple internal abscesses. The involvement of deer in the spread of bTB is held under constant review although it is accepted that wild deer are not a major risk when it comes to transmitting the disease to domestic livestock. To date, there has been no evidence of **brucellosis** or **rabies** in any wild deer in this country. **Rinderpest** has not been seen here since the nineteenth century, and whilst outbreaks of **anthrax** killed many deer in mainland Europe during two outbreaks in the 1960s and 1970s, it remains very rare.

Research that followed the 1967-68 outbreak of **foot-and-mouth disease (FMD)** proved that all of the British deer species were susceptible to the disease, at least under experimental conditions. In 1974 the government Animal Health Institute in Pirbright kept a number of deer in proximity to sheep with foot-and-mouth for two hours in a controlled experiment. The scientists found that all six native species of deer contracted the disease; reactions were especially severe in the deliberately infected roe and muntjac, whilst six of the nine muntjac used in the trial actually died.[3] However, it is noteworthy that none of them showed any signs of the typical lameness or increased salivation usually associated with FMD infection that would make it easy to recognise it in an animal 'on the hoof'. It was also found that, although the levels of the virus secreted were high enough to be a potential cause of infection in cattle, only in the droppings of some of the larger deer species did the virus persist long enough for them to be considered as carriers. There were no reported cases of FMD in deer during the major 2001 outbreak.

⬆ *(Top) Bovine tuberculosis (bTB) abscesses on the lungs of an infected deer.* **Note: such abscesses should not normally be cut open because of the risk of spreading infection**.

⬆ *(Bottom) Mesenteric lymph nodes swollen by bTB infection.*

➡ *Lesions in the mouth and sores between cleaves of hooves, both of which are strong indications of FMD infection.*

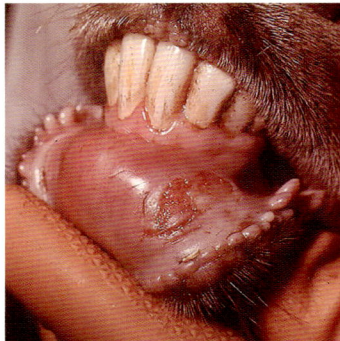

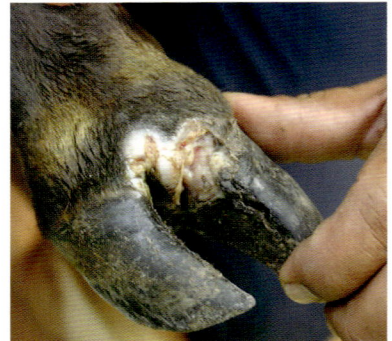

Chronic wasting disease (CWD) was first identified among wild deer in the USA during the 1960s and at the time of writing has yet to appear in the UK, although early isolated cases have been found in 2016 among reindeer and moose in Norway.[4] The disease is highly infectious and almost inevitably results in the death of the animal. It is, however, difficult to spot initially and it might take an infected animal up to two years to show clinical signs such as severe weight loss and change of character. Throughout this time, it will be shedding more and more infection. There are no known treatments. The prion responsible may live for up to fifteen years in some soils and easily resists most disinfectants, and while high concentrations of bleach have been found to eradicate it successfully, this is just not practical where clothing is concerned. The biggest risk is someone unwittingly bringing CWD back to the UK after a trip abroad, so biosecurity seems to be the key to keeping it out of this country.

Bluetongue is a viral disease which affects ruminants, primarily sheep, but can also infect deer. It is insect-borne, usually by midges, and is not contagious; it is not considered to be a threat to humans. The last outbreak in the UK was in 2007, but the risk is ever present of weather conditions being right for midges to carry the disease over from the continent.

All of the above are classified as notifiable diseases. It is worth repeating that, by law, any suspicion of their presence must be reported to the APHA who will arrange for all necessary testing and are an invaluable source of advice.

Photograph: Wyoming Game & Fish Department

↑ A North American elk with chronic wasting disease. Note the drooling, listlessness and emaciated condition.

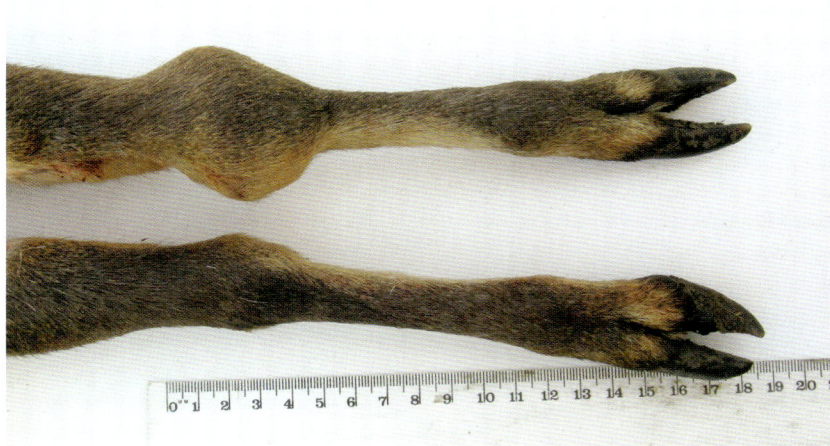

← Septic arthritis in the foreleg of a muntjac (top). Caused by a bacterium called Yersinia, it is often encountered around pheasant shoots where deer can pick it up from the bird's excretions. Though not a notifiable disease, any carcase showing signs of infection should not be passed into the food chain.

In general, there are often no external signs that something is wrong with an animal, beyond either unspecific wasting of the body or unusual behaviour. Loose faeces are associated with almost all serious illnesses in deer. The stalker should always be in the habit of checking a shot animal externally and internally for anything abnormal. FMD infection, for instance, often appears as sores between the cleaves of the feet or as ulcers in the animal's mouth. The best indicator of the need for a closer look is an examination of the lymphatic system, as described on page 172. However, I stress that if you see any signs that cause you particular concern, it is better to stand back and seek specialist advice rather than risk spreading any infection.

3 Gibbs, E.P.J., Herniman, K.A.J., Lawman, M.J.P. and Sellers, R.F. (1974), *Foot and Mouth Disease in British deer: Transmission of virus to cattle, sheep and deer*, Veterinary Record, 28 June 1974

4 Green, P. (2016), *Chronic Wasting Disease – the Norwegian cases*, Deer. Summer 2016

Main notifiable diseases of wild deer in the United Kingdom:

DISEASE	SYMPTOMS	NOTES
Anthrax	Infected animals generally die very quickly before any clinical signs are seen, but may exhibit shivering or twitching, a harsh dry cough, blood in droppings, fits and loss of appetite. Blood may eventually appear from all orifices; grossly enlarged spleen.	**Can be transmitted to humans** Spores can survive for many decades in soil etc. Never recorded among UK wild deer Legislation: Anthrax Order 1991
Bluetongue	The main signs are ulcers in the mouth, mucus/drooling from the mouth and nose, and swelling of the head and neck. Other signs can include reddened skin, fever, lameness or breathing problems. The characteristic darkened tongue comes in the later stages of the disease.	Not transmissible to humans Spread by midges so weather dependent; critical period is March – September Very rare among deer Legislation: Bluetongue Regulations 2008
Bovine tuberculosis (bTB)	Difficult to spot in a living animal, which may show signs of weight loss, loose faeces, poor appetite and coughing. Carcase inspection may reveal white or yellow lumps on the internal organs or attached to the body cavity; lymph nodes in the head/neck will be swollen.	**Can be transmitted to humans** Not common but occasionally noted in deer Legislation: Tuberculosis (England) (Amendment) Order 2014 & EU Directive 64/432/EEC
Brucellosis	Few clinical signs, but possibly swollen udders or testicles, nervousness or fever.	**Can be transmitted to humans** Last UK outbreak was among cattle in 2004; not identified in UK wild deer Legislation: Specified Diseases (Notification and Slaughter) Order 1992, Specified Diseases (Notification) Order 1996 & EU Council Directive 91/68
Chronic wasting disease	Highly contagious and fatal to the infected animal. Signs include separation from herds, increased thirst, drooling, teeth grinding or nervous or excited behaviour. Infected animals may exhibit poor co-ordination and stumbling, act listlessly or walk in repeated patterns. They may have tremors or paralysis and lose their fear of humans.	Not believed to be transmissible to humans No cases noted in the UK to date (2019)

DISEASE	SYMPTOMS	NOTES
Epizootic haemorrhagic disease	Difficult to identify unless the infection is severe, when the animal may show signs of fever, excessive salivation, lack of appetite, a swollen lining of the mouth and swollen red shin by the hooves. Easy to confuse with bluetongue.	Not transmissible to humans Never noted among wild deer in the UK Legislation: Specified Diseases (Notification and Slaughter) Order 1992 and Specified Diseases (Notification) Order 1996
Foot-and-mouth disease	Highly infectious. Sores and blisters on the feet, and ulcers in the mouth and on gums. Other signs include fever, shivering and lameness.	Not transmissible to humans Only rarely noted in wild deer Legislation: Foot and Mouth Disease (England) Order 2006
Rabies	Early signs include behavioural changes or hypersensitivity to light or noise. Later signs include unnatural or increased aggression, staring, itching and thirst. An animal in the final stages may froth at the mouth and have weakened muscles, show signs of paralysis, have convulsions and go into a coma-like state before death.	**Can be transmitted to humans** Eradicated in UK (apart from bats) in 1922. However, regularly recorded on the continent

A full list of notifiable diseases, along with comprehensive advice on how to spot and report them, can be found on the APHA website, and the Deer Initiative offer excellent best practice guidance both online and in hard copy.[5]

External Parasites

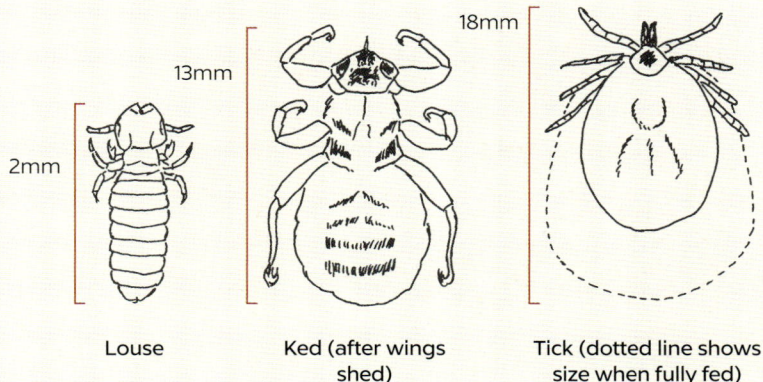

Louse, ked and tick

(Note that they are not drawn to the same scale)

13mm

18mm

2mm

Louse

Ked (after wings shed)

Tick (dotted line shows size when fully fed)

5 Deer Initiative Best Practice Guides, http://www.thedeerinitiative.co.uk/best_practice/

It is quite usual to find **deer keds** on an animal. They look like small, flattened, wingless flies with six legs that scuttle very quickly between the hairs of their host. About 13mm in length, keds initially have wings after pupating on the ground, but these are lost soon after they find a host and start to feed on blood from its skin. Keds are not always immediately apparent when a freshly-shot animal is inspected, but soon start to die and fall off as the carcase cools. Though unattractive to look at, keds carry no significant diseases and cause no damage to the deer beyond mild irritation. They are easily killed by modern insecticides.

Lice are only about 2mm in length and come in two main types – those that suck blood and those that feed on skin debris. They are wingless and may be first noticed as a reddish undercoat on the deer, particularly in the groin area. Very often the hair in the infected area is rubbed away. The shedding of the animal's winter coat in spring generally removes much of any louse burden.

Although lice have no real significance in the transmission of diseases, you should not expect to find them on every animal, and a heavy louse burden may well be a sign of poor condition that prompts closer examination for other problems.

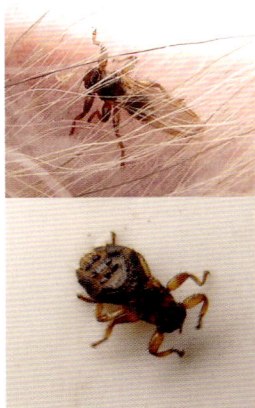

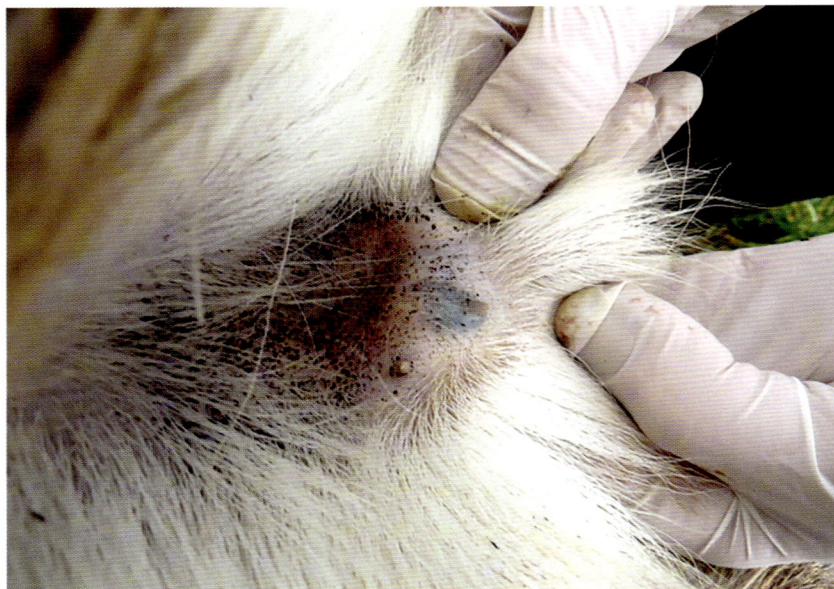

↑ *(Top) A deer ked that has recently landed on a deer. The wings will be shed and the parasite will live out the rest of its life-cycle on the host animal.*

↑ *(Bottom)) A ked once the wings have been shed.*

➡ *A heavy louse infestation of a Chinese water deer.*

⬇ *Ticks embedded in the groin area of a muntjac.*

Ticks are the most significant of the external parasites affecting deer. There are three main species, the most common of which is the castor bean sheep tick. It is quite usual to find them at various levels of engorgement, normally on the neck, belly, under the legs or behind the ears of an animal, with their heads buried in the skin where they feed on blood. They are particularly abundant in areas where sheep are farmed and where there is dense, warm ground cover such as bracken.

The tick life-cycle involves three hosts. The female tick feeds on Host 1 for about two weeks, then drops off onto the ground, where she lays her eggs. As much as a year later, the six-legged tick larva will emerge to feed off Host 2, normally a small rodent or bird, before falling off to moult. After remaining in the ground over winter, the nymph stage (now with a full adult complement of eight legs) eventually emerges and moults again to feed on Host 3 for about a week. It may then detach itself to spend as much as a year on the ground before achieving adulthood.

As ticks are blood-suckers and habitually move from host to host, they are capable of transmitting infectious diseases, although not all infected animals will show clinical symptoms and most ticks are disease-free. Of all the external parasites, it is the one of greatest concern to humans. The most important tick-borne illness is Lyme disease, and the tick is indeed the main vector for the bacterium that causes it. The classic first sign of infection is a red patch of inflammation around the site of the bite, which spreads over several days, although this may not be present in all cases. Other signs of human infection could include headache, general malaise and flu-like symptoms; eventually arthritis, meningitis or paralysis may occur. Although there have been fatal cases among humans, early treatment with antibiotics is fully effective. If you are bitten by a tick it makes sense to keep an eye on the bite and pay an immediate visit to a doctor if you have any concerns.

Anyone finding a tick attached to themselves should remove it with a special 'tick tool'. These are readily available and lift the tick carefully out of the skin without exerting any other pressure on it. Traditional remedies, such as tweezers or heat applied to the tick, are not recommended as they can encourage it to regurgitate infected blood back into the host.

The **warble fly** is more usually encountered in Scotland. The adult fly lays its eggs on the legs of the deer and the larvae, once hatched, work their way up to the animal's back. Once there they burrow into the skin and continue to develop, looking rather like oversized maggots, until they eventually burrow out again and fall to the ground to metamorphose into the adult fly.

Another is the **nasal bot fly**, which lays its eggs in the nasal passages of the deer. These hatch to become large maggots living in the nasal passages and throat. Affected animals can be seen shaking their heads and licking their noses to try to free themselves of the irritation. Nasal bot fly is also found north of the border, affecting both roe and red deer. It is becoming more common in England, where it also infects fallow deer. Neither it nor the warble fly can transfer to humans or affect the venison of the infested animal, although both conditions can be distressing to the animal and the latter will render skins useless if leather is required as a by-product from a carcase.

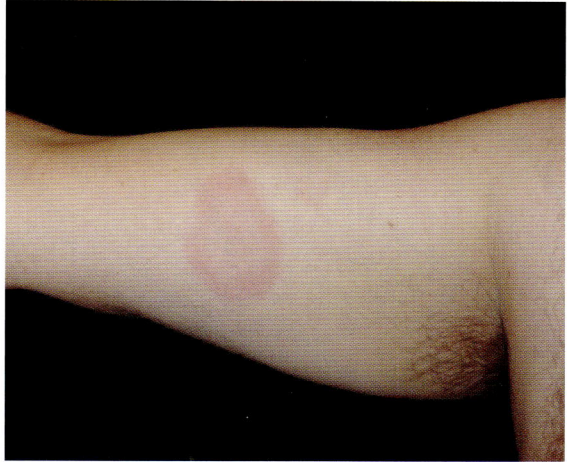

Photograph: Lyme Disease Action

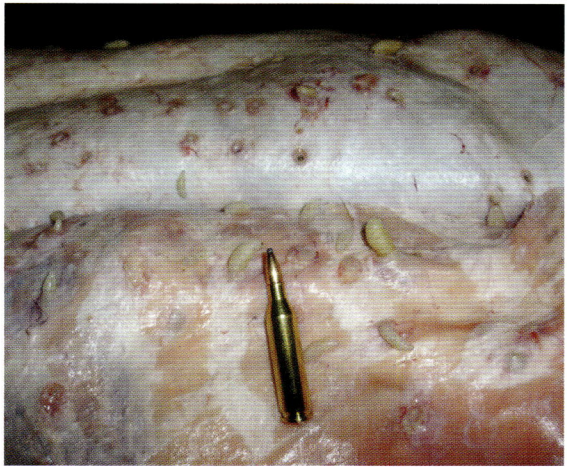

↑ (Top) A classic 'bullseye' rash caused by Lyme disease.

↑ (Bottom) Warble fly infestation on the inside of a red deer's skin.

Internal parasites

Moderate numbers of internal parasites are not unusual in deer, and in small numbers are of no real concern.

Liver fluke may appear as a mottled surface on the liver. When the liver is cut open, thickened tubes looking like clay-pipe stems and chalky deposits will be apparent, and the flukes themselves (looking like rolled-up leaves) can sometimes be seen in the bile ducts. The life-cycle of the liver fluke is complex and dependent on a specific species of snail (which favours marshy ground) to act as the secondary host. In deer living in well drained habitats it is uncommon, but in wet pastures and woodlands almost every deer may be affected. Fallow deer seem to suffer greater burdens of fluke than other species in the same area.

Shared grazing with cattle or sheep and wet, boggy pastures are the main risk factors for fluke in deer. Any liver found to be affected should be discarded, although the rest of the carcase will be fit for human consumption.

→ *The liver of a fallow deer, cut to show the flukes (arrowed).*

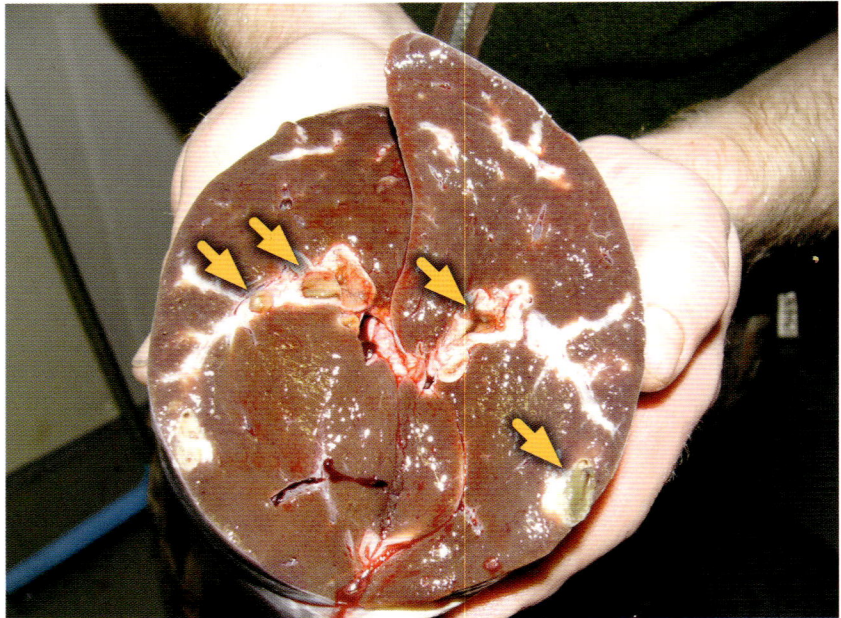

↓ *An individual liver fluke.*

Lung worms come in two types. Large lungworms (*Dictyocaulus*) resemble short pieces of thin cotton and are about 3 to 5 centimetres long. The exterior of an affected lung can have whitish patches or dead-looking grey areas. Severe infection with these worms is often seen in red deer grazing alongside affected cattle. Damage to the lung tissue from worm infestation can make it susceptible to invasion by pathogens, which will then contribute to the death of the deer.

An infested animal may cough as a result of the irritation in its lungs, or in an attempt to eject the worms in its airway; it may have a generally debilitated appearance and a scruffy coat.

The other type of lungworm (*Protostrongyles*) are much smaller and cause nodules in the lungs that feel like small hard peas or even like shot pellets embedded in the tissue. They are common in roe and fallow, but rarely have any ill effect. Some species of deer, most notably muntjac, are considered to be particularly resistant to lungworm infestation. Lungworm on its own will not affect the meat of an animal, so it is quite safe to eat, but if in any doubt it is wise to discard the lungs if they might otherwise be used (for instance, as dog food).

If you find a cyst, filled with clear liquid and about the size of a plum loosely attached within the body cavity of a deer (perhaps on the liver, gut or the wall of the abdominal cavity), chances are that it is the larval stage of a canine **tapeworm**. Occasionally such cysts can reach the size of a ping-pong ball. Before the life-cycle of the tapeworm can progress to the adult stage, the cyst must first be eaten by a carnivore such as a dog or a fox (tapeworm eggs are later passed in their faeces and picked up on vegetation by the feeding deer). Great care therefore needs to be taken when disposing of the gralloch, and dogs should not be fed any infected material. Only heavily infested deer carcases are considered unsuitable for human consumption.

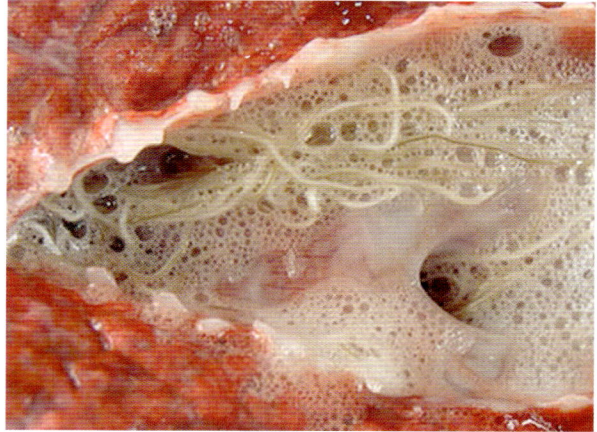

⬆ *An example of the lungworm* Dictyocaulus.

⬇ *A tapeworm cyst attached to the liver.*

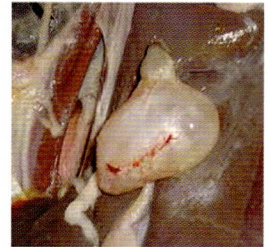

Suspect health – next steps

Assuming you have found something that looks suspicious in a carcase, what steps should you take next? Hopefully you know an experienced person whom you can consult for a second opinion, but perhaps you do not. In such a case your local farm animal vet practice is a good source of informed knowledge but you will probably have to take the samples to them. Storing a telephone number on your mobile phone will allow you to call ahead, and maybe even describe what you have found and have your mind put at rest. They will doubtless advise you on what to do should they be sufficiently worried by what you tell them. If a vet is not available, don't be afraid to approach the local APHA office directly; contact details are at Appendix II, Useful Addresses.

If you suspect a notifiable disease, you are legally required to move the carcase only as far as necessary to safeguard it from scavengers. Don't load it up and transport it away from the site.

When handling suspect organs, it is of the utmost importance to wear gloves – but you should routinely wear gloves for the gralloch anyway. The simplest approach is to put everything into clean plastic bags (a black bin bag kept in a pocket or your rucksack is ideal for any number of purposes). Smaller samples can be placed in freezer bags or the like, which you will hopefully be carrying anyway for keeping the liver and kidneys in as part of your normal gralloching routine. If you are planning to remove the stomach and intestines, a small puncture in the stomach wall will prevent it from ballooning up as bacterial action causes a build-up of gases.

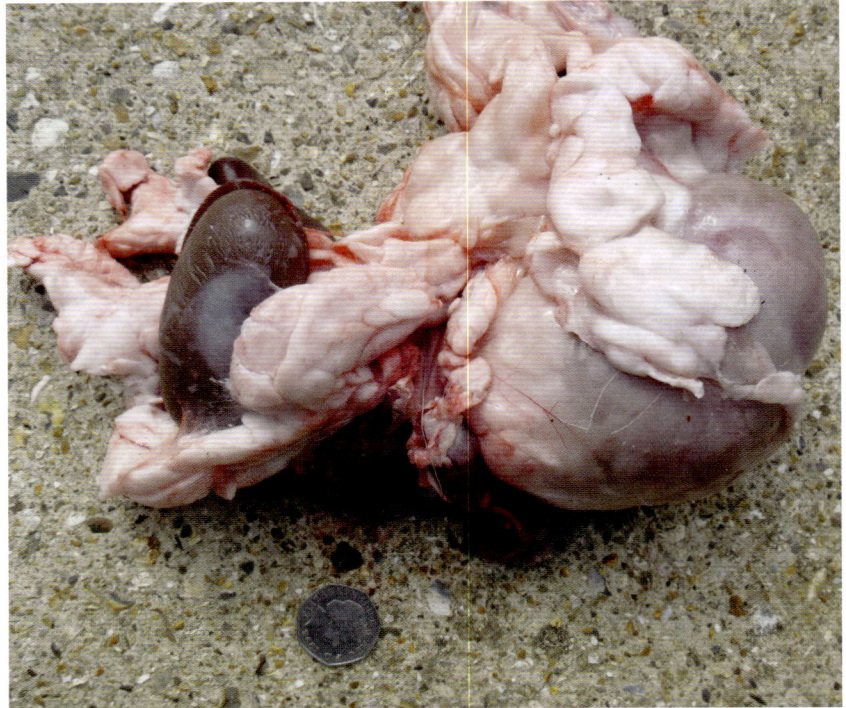

→ *Hydronephrosis in the kidney of a fallow deer (the left hand one is normal size), caused when urine is prevented by a blockage or obstruction to the bladder. While the kidneys should be discarded the rest of the carcase may safely pass into the food chain.*

If you do have to move the carcase, be careful to ensure that the site is clear before you leave it and mark the place where you found the deer. Suspect grallochs are best destroyed by incineration, but if you have to bury them do it effectively to prevent other wild creatures from finding and consuming it. Transport the carcase in a clean container, separate from any others that you may have shot. Take care that everything is kept cool, but do not freeze samples - this renders them unsuitable for examination and may kill off any micro-organisms that might otherwise show up. If you have to store things temporarily at home, make sure that they are not accessible to children or domestic pets.

Speed is of the essence when getting samples to a laboratory. If you are enlisting the assistance of your vet, they will have better facilities to prepare and package them. Otherwise, it is best to keep it simple. Glass or plastic pots and jars can be sterilised with boiling water, and clean plastic bags are an option for larger organs. Don't forget to label everything simply but clearly with the date, location, and any other details that might be pertinent.

Be prepared for a swift response from the APHA if your suspicions are confirmed by laboratory testing. A recent case of TB found in a roebuck resulted in a thorough decontamination of the site where it was shot, the vehicle in which it was transported and the stalker himself! And remember, if you can't manage to find the place where you shot the animal you are likely to look rather foolish.

You should not go away with the impression that such occurrences are everyday events for stalkers. They are not. Most of us will spend a lifetime without having to submit suspect samples for examination. However, as I hope I made clear at the beginning of this chapter, it is the duty of every deer manager to know what to do if anything out of the ordinary is discovered.

Human health

Apart from Lyme disease transmitted by ticks, a further word on another notable risk to human health posed by handling deer is important. While deer will carry parasites like many other wild animals, they are generally very healthy and as already discussed zoonotic diseases are rarely seen.

E. coli 0157, on the other hand, is present in the guts of deer as well as those of many other animals including sheep, cattle, goats, pets and wild birds, and carries a risk of very serious illness to young people and the elderly in particular. As far as deer managers are concerned, good personal hygiene is essential when handling carcases but as long as suitable precautions are taken and common-sense measures applied, any risks will be minimised. Routine use of disposable gloves, the frequent washing of hands with soap and water, application of antibacterial gel and care with handling snacks and flasks in the field should always be employed where appropriate. Contaminated meat, such as that tainted by stomach contents caused by misplaced bullets or poor carcase handling, is a special risk and should never be permitted to pass into the human food chain.

Occasionally, householders are worried by accumulations of deer faeces on garden lawns, especially where small children play. Small hands and mouths never seem to be kept apart for long, and this is a reasonable concern. In the longer term the only permanent solution is to exclude the deer from the garden altogether but the options available may not be feasible on account of cost or other considerations.

While the deer droppings should biodegrade fairly quickly, there are no simple solutions for cleaning them up and of course they will be quickly replaced if the deer continue to visit. Rainfall, or soaking with a garden sprinkler, will help to speed up the breakdown process. While picking the droppings up by hand is possible, this is both time-consuming and inefficient and it is strongly recommended that disposable plastic or latex gloves are worn at all times if you decide to do this. Alternatively, while using a rake or a brush is unlikely to be effective on grass, mowing the lawn should collect the majority of pellets along with the cuttings. Some people have even used a vacuum cleaner successfully on smaller areas. In the meantime, once again effective personal hygiene practices are essential under such circumstances, as they are after any physical contact with wild or domestic animals.

A few oddities

To round off this chapter, it might be useful to mention a few oddities that you may come across from time to time. Deer are frequently injured fighting among themselves, in road traffic accidents or by other means, and it is astonishing at times that some animals can lead normal lives after recovering from quite serious injuries. Some animals can even lose a limb and still move about normally, running at much their normal speed on only three legs and easily keeping up with other members of their herd. Males have been known to rut successfully, while females have given birth to and successfully raised their young to maturity. Occasionally, genuine 'freaks' appear, such as animals with two fully developed hooves at the end of one leg: a condition known as 'polydactylism', which has been noted in a number of species. I am also aware of a muntjac buck which never grew a tail - very much a one-off, as I have never heard of another instance.

➡ *The forefeet of a fallow deer with polydactylism.*

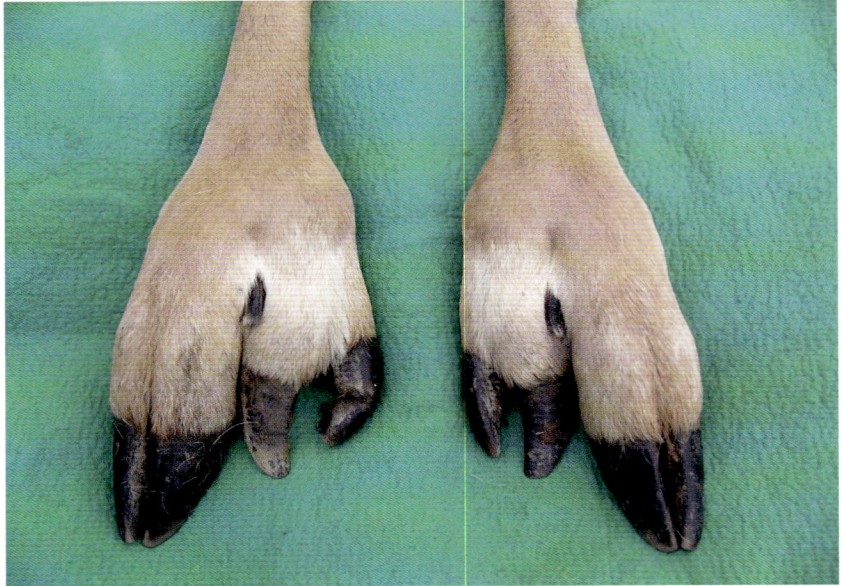

Scars are commonplace on mature muntjac and Chinese water deer bucks that have received them during fights over territory, as are ripped ears from the same causes. The sharp canine teeth are capable of doing far greater damage to an opponent than the rudimentary, in-curved antlers of the muntjac, which tend to be employed more in shoving matches when a trial of strength occurs. The slashing canines of a muntjac or water deer buck are used quickly and accurately, and are not to be trifled with. A mature buck that is not carrying any obvious injuries from such altercations is usually either a more dominant specimen, or an indication of low population densities with less frequent clashes taking place. Broken canines are commonplace by middle age and are not replaced, much to the disadvantage of the owner. Broken antlers are at least regrown after casting.

Antler malformations can be more commonplace than you might suppose. The growing antler is effectively living bone, which dies and hardens off at the end of the growth cycle, and until that time it is particularly sensitive to any number of mishaps. Damage to the pedicle, the bony protuberance on top of the skull from which the antler grows, can cause the antler to grow out of it at a bizarre angle. Other malformations may be attributable to knocks and bumps, dietary deficiencies, disease or unusually high infestations of internal parasites.

⬇ *Ripped ears and other scars are common among fighting muntjac bucks.*

Roe deer seem particularly prone to antler malformations, which is probably not surprising given that this species, unlike the others, regrows its antlers throughout the winter when there will be increased disturbance from game shooting and other activities. Fleeing animals are more likely to damage accidentally the growing tissue of their regenerating antlers, causing them to produce unusual shapes and sizes.

Hummelism is a condition more commonly seen among red deer, but also very rarely in some others, where the adult male fails to produce antlers at all. As hummelism is usually linked to failure of the growing animal to reach a threshold body weight before it can develop the pedicles from which the antlers will eventually grow, it is unlikely to affect the smaller species, which do not have so far to go before they attain adult body weights. There is no evidence that the condition is hereditary although most managers prefer to remove such animals from the herd.

Campylognathie, or bent-nosed syndrome, has been noted in several species. It is a rare phenomenon that results in the nose bones growing at a twisted angle. There is as yet no scientific explanation for the condition, but it may be genetic. Some have suggested that the malformation follows the laws of coriosis, which also dictate which direction water swirls in when it goes down a plug-hole, but this does not appear to bear true for many specimens.

Generally speaking, white deer are very uncommon and only the fallow naturally occurs in a white variety. Fallow come in a number of colours, most usually described as common, menil, black and white, although there can be some variations between these main colour phases. The white variety of fallow is quite normal and is not in fact a true albino. Albino deer have pink skin and pink irises to their eyes; white fallow have normal eye pigmentation, although their hooves and noses might be somewhat paler than the other colour varieties.

Black or melanistic fallow are very common, and indeed seem to be the dominant fallow colour variety in some areas. Melanism is caused by an over-production of a chemical called melanin, which causes darker pigmentation in animals. Albinism is caused by a lack of melanin. White deer of any kind stand out easily to predators against most backgrounds, and it is for this reason that wild deer managers are split in their attitude towards whether or not to preserve them. On one hand, their presence makes it easier to spot herds containing more effectively camouflaged animals, whilst on the other their especially high visibility can attract poachers and other unwanted attention.

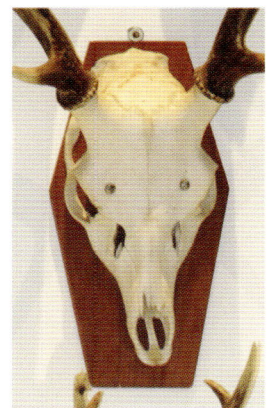

Photograph: Valerie Masterton

→ *A piebald roe buck photographed in Yorkshire. Such animals are only rarely encountered.*

Occasionally, white deer are seen amongst other species. White red deer are sometimes seen in the wild, and a few exist in park herds. White sika have been regularly observed in recent years in the Purbeck area of Dorset, and white roe are also occasionally seen. Very often, though, these animals are not actually pure white but simply paler than the normal coat colour. Piebald deer have occasionally been recorded, most often roe but occasionally muntjac as well.

→ *White 'socks' on a muntjac.*

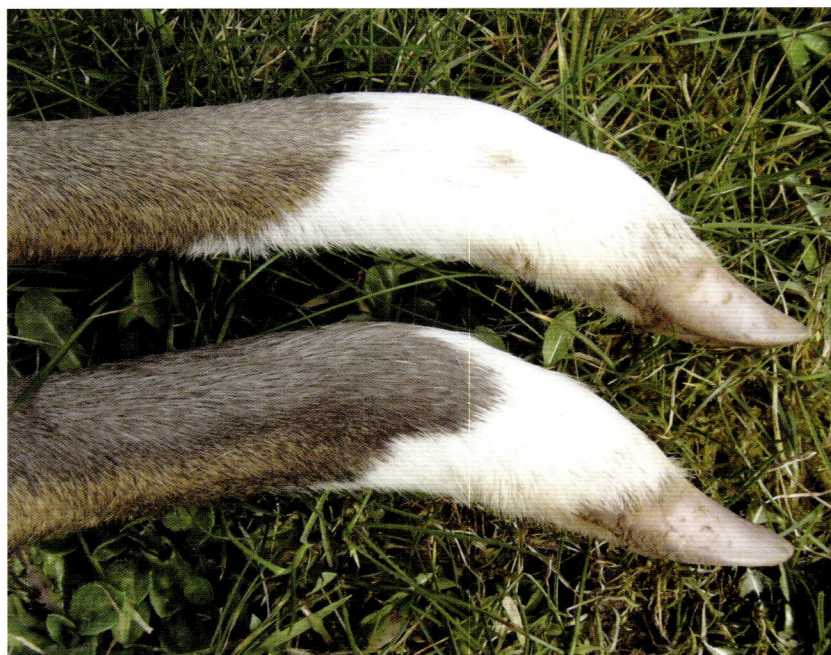

The specially adapted digestive tracts of deer contain micro-organisms which are specifically capable of destroying the cellulose walls of plants, a process which is further assisted by the act of rumination – the regurgitation of food matter for further chewing and breaking down prior to being swallowed again.

Deer are occasionally encountered showing signs of diarrhoea such as badly soiled rump areas. While this might be a side-effect of an infection, as deer are generally very healthy this is not usually the most likely explanation. In spring, diarrhoea is very commonly caused by too rich a diet with too little fibre as a result of the flush of new growth and buds, which the animal's digestive system fails to deal with efficiently.

At other times, diarrhoea could be put down to a sudden change of diet, as the micro-organisms in a deer's stomach, essential to digestion, may take time to adapt to something they are not used to. Indeed, it may simply be that the deer has eaten an unsuitable toxic plant which has had this temporary effect. There is generally little need for concern and in any case there are not really any practical ways to intervene with a wild deer, whose digestive system should return to normal in due course. Dehydration is usually not an issue as deer generally obtain all the moisture they need though their foodstuffs, although if this proves insufficient they will drink water.

While deer are most certainly herbivores, they are not averse to ingesting unusual objects and some strange things have been found in the stomachs of dead deer. These include polythene bags, baler twine, fired cartridge cases and even condoms. In Richmond Park alone, eating litter is believed to be responsible for the deaths of around five deer every year. Bezoars occasionally form in the stomachs of some animals, including deer, when an indigestible object becomes stuck inside the digestive tract and over time becomes coated with minerals until it becomes smooth and hard. In many ways bezoars are a sort of mammalian pearl, having parallels with what happens when a foreign object becomes trapped inside a mollusc.

Although meat is not a food source normally sought by deer, it would not be strictly true to say that they never eat it. Although many species are known regularly to chew old bones or antlers, red deer in particular have been recorded as killing the chicks of the Manx shearwater on the island of Rhum by biting off and swallowing their heads.[6] There have also been reports of stags eating grouse chicks, rabbits killed in snares, and even swallowing ducklings in parks. Such behaviour probably has much to do with a craving for calcium among male deer that are growing new antlers, and is certainly not related to regular sources of nutrition. ■

6 Brooke, M. (2010), *The Manx Shearwater*, Poyser Monographs, London

10 The Law and Deer

Although deer receive some important protection under the law, this protection is largely driven by the need for humane treatment and it is important that everyone involved in their management – whether lethal or otherwise – understands what is and is not permitted. Other aspects linked to such management, whether it be the possession and use of firearms, building associated structures or even passing legally culled venison into the human food chain, are covered by relevant regulations.

Some laws have been alluded to already at the relevant point in previous chapters. Overall regulation is a huge subject which cannot be covered comprehensively here, so what follows is not exhaustive: it should only be taken as a working guide towards all the many legal aspects of dealing with deer and should not be regarded as definitive. Furthermore, it is important to be aware that the law can vary between England and Wales, Scotland and Northern Ireland, each of which has its own specific legislation. Laws are always subject to change too, so it pays to remain abreast of developments.

Some further reading on the subject is suggested at Appendix I.

Primary deer legislation

In England and Wales, the main legislation specifically concerning deer is the Deer Act of 1991, which updates the original 1963 Act. In Scotland it is the Deer (Scotland) Act 1996, and for Northern Ireland, Articles 19 to 23 of the Wildlife (Northern Ireland) Order 1985. All cover such aspects as close seasons, permitted firearms and specific offences relating to the treatment of wild deer.

Do not be misled by the dates as there may also be later amendments. For example, a 2007 amendment to the Deer Act 1991 updated (among other things) the close seasons for female deer, the legality of shooting from vehicles and permitted .22 centrefires to be used for shooting muntjac and Chinese water deer.

↓ Deer legislation details permissible means of controlling wild deer; under most circumstances an appropriate calibre of rifle is specified.

Statutory bodies

Much of the law as it applies to deer is administered at a more regional level by bodies set up by government to consider issues and make judgements as appropriate. These statutory bodies are Natural

England, Scottish Natural Heritage, Natural Resources Wales (on behalf of the Welsh government) and the Department of Agriculture, Environment and Rural Affairs Northern Ireland. These should be taken as first points of contact when applying for licences and so on; contact details are listed at Appendix II. Responsibility for the actual enforcement of much of the legislation rests with the police.

Some older sources of information regarding Scottish deer matters may refer to the Red Deer Commission, which ceased to exist in 1999, or its replacement, the Deer Commission for Scotland, which itself merged with Scottish Natural Heritage in 2010.

Ownership

Nobody owns wild deer. The law defines them as *ferae naturae*, wild animals, and *res nullius*, or things that have no owner. It follows that as long as they are wild (i.e. free to roam where they please) they cannot be stolen and only become property once they have passed into somebody's possession. It also follows that nobody can be held responsible for the behaviour of wild deer, for example if they are involved in collisions with road vehicles or cause damage to a neighbouring farmer's fields. A degree of liability for the damage, or potential damage, caused by deer does however exist in Scotland (see below).

Only once the deer has been 'reduced into possession' (the technical phrase), can it be considered to be someone's property. For this to be legal, the person who does this must have the right to kill the animal, normally through holding the sporting rights on the land in question. This might be the landowner, someone acting on their behalf or someone who has been given the sporting rights. Although an occupier or tenant of land in England and Wales has certain rights to take action when deer are raiding their crops, they still may not do so without the authority of

➡ *A live wild deer belongs to nobody. It only becomes property once it has been legally reduced into possession.*

whoever holds the shooting rights; only in Scotland has the tenant any independent right to take action against marauding deer.

You occasionally hear the old rural myth that a person who hits a deer or game bird with their car may not pick it up, but that the driver of the car behind may legally do so. This is not true. As a simple rule of thumb, the carcase belongs to the owner of the land on which it falls, much in the same way that a pheasant, shot dead whilst in the air, technically belongs to a neighbouring landowner if it reaches the ground on their side of the boundary. While some may argue that they have a moral right to recover a shot carcase, there are of course the matters of trespass and ownership to consider, so it makes good sense to come to a suitable agreement with neighbours to avoid unnecessary disputes. Once legally collected, however, the carcase remains the property of the shooter whilst it is in transit. Retrieval of a shot deer from over your boundary is catered for in law but only if certain conditions are met.

Deer living on farms, in parks or other collections, kept under clearly enclosed circumstances, are looked at differently by the law and treated as property even when still alive – and even if they are escapees subsequently encountered elsewhere.

Liability

Under limited circumstances, if the tenant of an agricultural holding suffers damage caused by deer that he does not have the right to cull, a right to apply for compensation exists. A claim can be made under the Agricultural Holdings Act 1986 or the Agricultural Holdings (Scotland) Act 1991, but written notice must be given to the landlord by the tenant within one month of the damage becoming evident to allow for an appropriate inspection to be made. It is possible for the landlord to be indemnified if the shooting rights are held by a third party, such as a stalking tenant or shooting syndicate.

Only in Scotland is there a wider implied liability for the activities of wild deer. Scottish Natural Heritage has the legal power to enact mandatory Control Schemes if a voluntary approach has not been agreed to address situations where there is crop or woodland damage, competition with livestock, a danger to the public or simply a perceived need to reduce deer numbers. Such action may also be taken to protect natural heritage interests, and SNH staff or anyone authorised by them in writing may enter the land, either to carry out a census or to control operations. Furthermore, SNH is entitled to sell the carcases of any deer shot under mandatory Control Schemes and charge the landowner for the balance of any expenses incurred.[1]

There are no similar powers in England, Wales or Northern Ireland.

Seasons

Until just over fifty years ago, there were no close seasons to protect deer from persecution in this country, and it was primarily the tireless efforts of what was to become the British Deer Society that eventually afforded them protection from being treated as vermin with no consideration for breeding seasons or humane treatment. One could say that a similar situation exists at the present time with wild boar in the UK. The primary deer legislation introduced close seasons and restrictions on when deer may be shot, which tend to revolve around the critical

1 Deer (Scotland) Act 1996, Sections 8-13

breeding times for female deer and antler growth for males. At the present time (2019) the **open** seasons, i.e. when deer may be taken by legal means, are:

Open season, United Kingdom			
	ENGLAND AND WALES	**SCOTLAND**	**NORTHERN IRELAND**
Red and sika males*	1 August – 30 April	1 July – 20 October	1 August – 30 April
Red and sika females*	1 November – 31 March	21 October – 15 February	1 November – 31 March
Fallow males	1 August – 30 April	1 August – 30 April	1 August – 30 April
Fallow females	1 November – 31 March	21 October – 15 February	1 November – 31 March
Roe males	1 April – 31 October	1 April – 20 October	*Not present*
Roe females	1 November – 31 March	21 October – 31 March	*Not present*
Chinese water deer (both sexes)	1 November – 31 March	*Not present*	*Not present*
Muntjac (both sexes)	No close season		

* includes hybrids

Because it is so difficult to differentiate between male and female Chinese water deer in the field - this species does not carry antlers and has no other readily distinguishing features between the sexes - the open and close seasons for them are the same. The muntjac, on account of having no fixed times of year for breeding, cannot be afforded the deliberate protection of a close season although it is recommended that when culling female muntjac, only immature or heavily pregnant does are selected to avoid leaving dependent young.

➡ *Shooting seasons afford protection against the inadvertent orphaning of dependent young.*

Time of day

Deer may normally only be shot between one hour before sunrise and one hour after sunset, wherever you may be in the United Kingdom.

Night shooting requires a licence from one of the national statutory bodies and can be very difficult to obtain in some places. Good reason will need to be proven before the granting of a licence will be considered: usually this will relate to public health, public safety, conserving the natural heritage or preventing serious damage to property.

↑ *Under normal circumstances deer may only be shot from one hour before sunrise until one hour after sunset.*

Public holidays

As long as they are in season, deer may be legally shot on any day of the week, including public holidays in England and Wales, Scotland or Northern Ireland. However, it is not customary to do so in Scotland on Sundays or on Christmas Day.

Legal calibres

The Deer Act and other relevant legislation is also very specific about what may be used to kill deer under normal circumstances. One of the driving forces behind the introduction of modern measures was the way in which deer were killed indiscriminately with whatever firearm was to hand, irrespective of its suitability for the task. As a result, many more animals were probably wounded rather than killed outright. For the shooting of healthy deer under normal circumstances today the following are the minimum requirements for rifles:

Minimum rifle requirements, United Kingdom

	CALIBRE	MUZZLE ENERGY	MUZZLE VELOCITY	BULLET WEIGHT
ENGLAND AND WALES				
All deer	.240"	1,700 ft/lbs	N/A	N/A
Muntjac and CWD only	.220"	1,000 ft/lbs	N/A	50 grains
SCOTLAND				
All deer	N/A	1,750 ft/lbs	2,450 fps	100 grains
Roe deer <u>only</u>	N/A	1,000 ft/lbs	2,450 fps	50 grains
NORTHERN IRELAND				
All deer	.236"	1,700 ft/lbs	N/A	100 grains

↑ *A range of appropriate ammunition for deer management. From l to r, .22 (for comparative purposes only), .243 Winchester, .308 Winchester, 6.5x55 Swedish, .270 Winchester, 30-06.*

Of the main commercially available firearms calibres, the .243 Winchester with an appropriate weight of bullet is probably the most practical minimum calibre for shooting all species of deer in the UK. No maximum calibres are laid down, although choice may be limited by what the police licensing authorities consider appropriate.

For muntjac and Chinese water deer in England and Wales and roe in Scotland, many (but not necessarily all) of the .22 centrefire calibres are suitable - although great care must be taken in the selection of ammunition in order to ensure that it meets the minimum legal requirement.

In all cases, the bullets used must be designed to expand on impact (although the precise wording of this requirement may vary between different pieces of legislation). This is to ensure that sufficient energy is transferred to give a reasonable expectation of a humanely killed animal. A solid bullet which does not deform properly may not do so and neither would a bullet which expands too quickly on impact and fails to penetrate effectively to the vital organs.

Under very limited circumstances a shotgun may also be permitted for shooting healthy and uninjured deer (see below).

Damage prevention

In England, Wales and Northern Ireland it is permissible, on any cultivated land, pasture or enclosed woodland, for an authorised person (normally the occupier of the land or a person in their employ) to kill or take deer out of season if damage to

→ *The use of shotguns for shooting healthy wild deer is restricted for good reason. A pellet, probably from a light game shooting load, can be seen embedded in the pedicle of this muntjac buck which had subsequently recovered.*

crops, vegetables, fruit or growing timber is taking place. To justify this, it is necessary to demonstrate that there is reason to believe that deer of the same species are responsible, that it is likely that further serious damage will be caused and that the action was essential to prevent such damage. This is one of the occasions where a shotgun may be used, although it must not be less than 12 bore, and the shot size AAA only.[2] In Northern Ireland, a non-spherical projectile of not less than 350 grains is also allowed.[3]

Similar provisions are available in Scotland for the protection of crops. Once again, a shotgun is permitted, but it must fire a single rifled non-spherical slug of not less than 380 grains in weight or a cartridge loaded with SSG shot or larger (for roe deer AAA or larger is allowed).[4]

The circumstances which permit the above actions can sometimes be complicated, and it is recommended that specialist advice is sought before they are taken.

Firearms legislation

The law concerning firearms ownership and use in Britain is primarily covered by the Firearms Act 1968, although there has been a great deal of subsequent legislation to update this. In general terms, it is a very serious offence to possess, purchase, hire, borrow or otherwise acquire firearms or their ammunition without a relevant licence.

A firearm is defined by law as a 'lethal barrelled' weapon capable of causing injuries from which death may result. The definition also includes any component part of such a weapon, such as the barrel, bolt or action, and it follows that such a component part has to be treated as a firearm in its own right. Sound moderators fitted to deer calibre rifles are also currently treated as firearms in their own right and must be licensed and kept in the same way. As an aside, although the Firearms Act frequently uses the term 'weapon' it is far preferable to say firearms. 'Weapon' implies to the modern ear anything which might be used to harm people, whereas a firearm is no more than a working tool.

Broadly speaking and without going into very detailed definitions, there are three main classes of firearm which are relevant, as far as deer management is concerned:

Section 1 - Sporting rifles, and shotguns with magazines capable of holding more than two rounds of ammunition. These must be held on a Firearms Certificate (FAC) issued by the police licensing authorities. The FAC must specify the type and calibre of a desired firearm before the certificate holder can actually acquire it, and will also strictly specify aspects of its use such as quarry species, ammunition holding and where it may be used.

Section 2 - Sporting smooth bore shotguns (other than above). These are held on a Shotgun Certificate (SGC), which is a more general document than a FAC. The holder may use it to acquire any Section 2 firearm without the need for prior permission. There are generally no conditions on a SGC specifying actual use provided that all is within the law, although there will be specific requirements for security, transfer etc.

2 Deer Act 1991, Section 7

3 Wildlife (Northern Ireland) Order 1985, Article 20 (6)

4 Deer (Scotland) Act 1996, Section 26

Section 5 - 'Prohibited weapons' – such as handguns, semi-automatic full-bore rifles etc. These are also held on a FAC but are subject to far more stringent controls regarding their ownership and use, and an applicant will need to demonstrate very special circumstances before the licensing authorities will grant permission to hold one.

In Northern Ireland, a FAC is required for all firearms and ammunition, including smooth bore weapons such as shotguns and air rifles.

➤ A Section 1 rifle is the normal tool for deer management. This one is fitted with a sound moderator.

As already stated, a sound moderator assumes the characteristics of the firearm to which it is fitted, and must be held on a certificate as appropriate.

(Note: Each 'Section' referred to above means the relevant section of the Firearms Act, which defines the type of firearm in fuller detail. There are no Section 3 or 4 firearms as these sections cover other matters.)

The procedure for acquiring a FAC or SGC is straightforward. Most licensing authorities have the relevant forms available for download on their parent constabulary website, including instructions on how they should be completed along with details of fees, referees required and so on. Before a licence is granted you should be prepared for a visit from a representative of the licensing authority to confirm your suitability and to check your arrangements for storage and security. These people, usually known as Firearms Liaison Officers, are an excellent source of advice in the first instance, and it often pays to contact their department to discuss your needs and plans before submitting an application. The applicant will need to be able to provide the licensing authority with good reason for wanting to acquire and keep a firearm; proof of permission from a landowner or any other person holding the sporting rights for any ground in question also needs to be available for examination. In some cases, the licensing authority may wish to physically inspect the ground to check that it is suitable for the firearm requested.

Capturing, moving and releasing deer

The capture and movement of live wild deer, usually with the aid of traps, nets or stupefying drugs, requires a licence from the appropriate statutory body before such a task can be legally carried out. Such procedures demand specialist knowledge and experience and should not be undertaken lightly, although sometimes it is necessary to move deer or capture them alive for educational or scientific purposes. Any drugs used will be subject to strict control, as will the dart guns or blow pipes required to deliver them. In all cases it is advisable to seek specialist advice before proceeding further.

The release of deer is also strictly controlled. While licences apply to their capture, the native UK deer species (red, roe, and also fallow, which are treated as native by the law) can be legally released anywhere in the country without prior permission although it should be noted that roe are not considered to be native in Northern Ireland.

For muntjac, Chinese water deer and sika the situation is different. Under the Wildlife and Countryside Act 1981,[5] it is illegal to either release or allow these species to escape anywhere in the UK – although if a licence has already been issued for tagging or collaring, the condition for release afterwards will be included on it. There is a specific prohibition on the release of sika, or indeed any deer of the genus *Cervus* or their hybrids, on the Outer Hebrides, Arran, Islay, Jura and Rum. This is a specific measure to protect the genetic integrity of Scottish red deer.

Until now, Natural England has been able to issue a licence for the release of muntjac which had been rehabilitated after being injured, such licences usually only being granted to wildlife rescue centres and official organisations. At the time of writing, however, it is understood that the Invasive Alien Species (Enforcement and Permitting) Order 2019 will supersede the provisions of the Wildlife and Countryside Act in England and Wales. This means that no more licences will be issued to release muntjac under any circumstances, nor to keep them in captivity, apart from in the most exceptional conditions.

Otherwise, under existing law, it is illegal to release muntjac into the wild anywhere – even if the animal has inadvertently become trapped in an enclosure such as a pheasant release pen. It is reasonable to assume that the recent arrival of the muntjac in Northern Ireland, as well as the rumoured presence of roe deer, can be attributed entirely to illegal human agency.

It is worth noting that under the Wildlife and Countryside Act 1981, it is illegal to use any live animal for the purpose of taking or killing any deer.

Unlawful weapons

Traps, snares, poisons and stupefying baits or nets may not be used to take any deer; in fact it is an offence even to set any of these deliberately in such a position that might injure a deer that comes into contact with them.

The Deer Act 1991 specifically prohibits arrows, spears or other missiles, whether discharged by a gun or otherwise, containing poisons, drugs or muscle relaxants to take deer (although the use of suitably licensed darting equipment is permissible under special circumstances).

5 Wildlife and Countryside Act 1981, Schedule 9

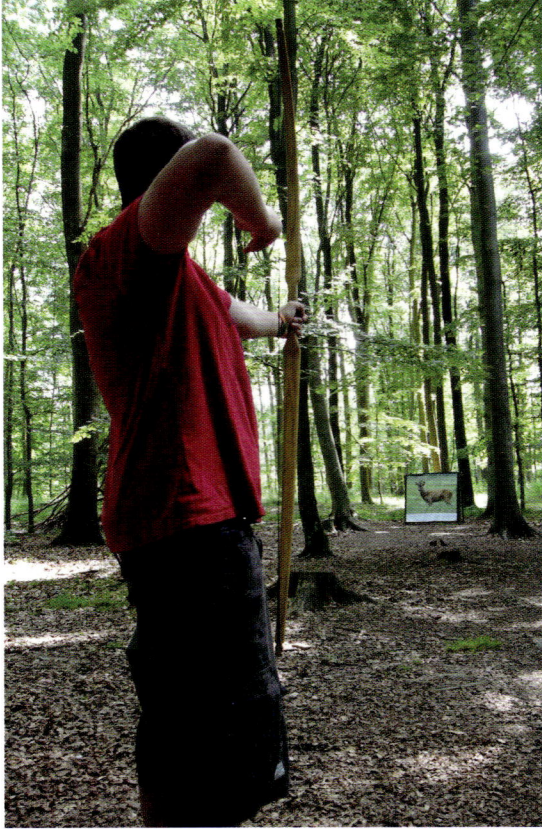

↑ While it may be legal in other countries, the bow hunting of deer or any other wild animal is prohibited in the UK.

Additionally, using a bow or crossbow to kill or take *any* animal is prohibited under the Wildlife and Countryside Act 1981.

Vehicles

Though there are variations in the way that the law is worded in different parts of the UK, in general terms it is permissible to shoot deer from a vehicle. However, the vehicle must not be moving and the engine needs to be turned off. It is not permissible to use a vehicle to drive deer to guns in culling operations.

Scottish legislation does allow for the use of helicopters to assess culling areas, move teams of rifles about, co-ordinate operations or retrieve carcases. They may also be used to move deer for the purpose of counting them or moving them away from areas where they are causing damage or pose a threat to safety. However, moving deer with the intention of killing them remains unlawful.

Dogs and deer

While the deliberate hunting of deer with dogs was made illegal by the Hunting Act 2004, an important provision for the conscientious deer manager is the allowance within the same Act for no more than two dogs to be used to hunt for an injured deer. This is vitally important to anyone responsible for shooting deer, as occasionally a shot may go wrong and a trained dog enables a wounded animal to be found and despatched without undue delay. It is also permissible to use a dog to accompany you when stalking deer, a practice that many find useful as the dog's superior senses will often locate deer that the human may not have spotted.

It is worth making clear at this point that uncontrolled dogs seen in pursuit of wild deer are not subject to the same rules that allow a farmer to shoot them if chasing livestock.

Animal welfare and humane despatch

The Wild Mammals Protection Act 1995 sets out to protect wild animals from deliberate cruelty and all who are involved in the humane despatch of injured animals, for instance after a collision with a vehicle, need to be aware of it. It is illegal for any person to 'cruelly kick, beat, stab, impale, burn, crush or drown any wild mammal'. Exceptions to this rule apply if it is possible to show that the attempted killing was an act of mercy after the animal was severely disabled, that there was no reasonable chance of it recovering and that the process was conducted in a reasonably swift and humane manner.

It should be noted that the Animal Welfare Act 2006 does not apply to wild deer and covers only commonly domesticated animals not living in a wild state. It is only likely to apply to farmed deer, although tame, captured or injured wild deer under treatment also fall within its provisions.[6]

As far as the humane despatch of an injured deer is concerned, many of the laws that apply to uninjured deer do not apply. The 2007 amendment to the Deer Act permits a person to use 'any reasonable means for the purpose of killing any deer if he reasonably believes that the deer has been seriously injured, other than by his unlawful act, or is in such a condition, that to kill it is an act of mercy'.[7] In practical terms, this means that any firearm and ammunition combination, irrespective of whether it would be legal for shooting a healthy deer, may be employed, as might other means of ensuring a rapid loss of consciousness and death. Beware, though, that you may breach FAC conditions relating to the use of non-legal deer calibres for the purpose if you do not have the provision to use that particular rifle for humane despatch. So, while you may not be committing an offence under deer legislation, you may be doing so under the entirely separate Firearms Act.

A word of caution is also appropriate at this point. The efficient and humane euthanising of a stricken animal may still appear cruel to onlookers, so anyone conducting humane despatch is strongly advised to remain aware of public sensibilities at all times. There have been several cases of cruelty accusations made in the past against individuals acting with the best of intentions.

↑ *These farmed red deer, clearly ear-tagged, are subject to different legislation from that which applies to their wild counterparts.*

6 British Deer Society (2017), *Training Manual for Deer Stalkers*, Fordingbridge, BDS

7 Regulatory Reform (Deer) (England and Wales) Order 2007, Section 6 (4)

↑ *This roe buck was blind in one eye and virtually immobilised by serious internal injuries, probably as a result of a vehicle collision. The decision to cull such an animal, even if out of season, is clearly justifiable on humane grounds.*

Health and safety

As in all walks of life, health and safety regulations apply to activities involving the management of deer. The main piece of health and safety legislation and regulation forming the backbone of requirements is the Health and Safety at Work etc Act 1974 (as it is officially designated); the Act exists to ensure practical compliance and helps everyone to understand and implement relevant health and safety legislation. Other pieces of legislation and regulation also exist to cover a wide range of industries.

Whilst primarily protecting the rights and general health of employees, and complicated by its size and complexity, much of the legislation is common sense and can be satisfied by care, attention and appropriate use of risk assessments. Deer management, which may involve the use of firearms, high seats, chainsaws and other equipment, can be considered a higher risk activity and it is important that due care is taken in appropriate areas. Although purely recreational stalkers would tend not to fall within and be subject to health and safety regulations, anyone actually conducting deer management activities on behalf of a landowner might be considered to be 'conducting an undertaking', and thus incur a responsibility for ensuring that others are not put at risk.

The Health and Safety Executive is an important source of advice and guidance, which can be accessed through the relevant website.[8]

8 Health and Safety Executive website; available at www.hse.gov.uk (Accessed 30 August 2018)

Game meat hygiene and waste disposal

Under the Food Safety Act 1990, anyone involved in supplying venison has a liability for ensuring that it is fit for human consumption. This liability includes estates and individual stalkers and also extends to all other aspects of the supply chain, such as game dealers, restaurants and retail outlets. In essence it ensures that all food meets food safety requirements, is not injurious to health, is not falsely or misleadingly described and is of the proper quality or nature required. The penalties for offences under the Act are potentially very severe.

Suppliers of venison into the food chain (the term 'supply' includes barter and exchange as well as sale) are responsible for appropriate levels of hygiene and record-keeping. Furthermore, anyone who hunts wild game of any kind with a view to passing it into the food chain should have 'trained hunter' status, possessing a suitable level of knowledge of pathology and processes to make an initial inspection of the carcase as soon at it has been killed. Not every member of a hunting team needs this status, but at least one person does and will be held responsible for making a declaration that the relevant checks have been carried out and noting any particular abnormalities on a numbered tag which accompanies the carcase. Training must be conducted by a recognised authority (for deer stalkers, the Deer Stalking Certificate 1 provides such a proof of competence).

Wild animals that have died of natural causes do not need to be disposed of in any specific way. However, under the Animal By-Products Regulations 2005, deer carcases, gralloch or larder waste must be incinerated if the animal is diseased, the product of meat processing or used to produce game trophies.

Trespass

The law regarding trespass can be complicated and is frequently greatly misunderstood, so a few words on the subject may be helpful. Matters are not helped by differing interpretations of rights of way and the 'right to roam', as well as some of the myths which have grown up around the latter.

The laws of trespass were first created many hundreds of years ago to protect the king's peace, and have evolved considerably since. In simple terms, trespass can be defined in modern terms as 'an unlawful intrusion which interferes with one's person or property'. It is more commonly taken to include acts of intentional and wrongful physical invasion of someone else's property, such as unauthorised entry onto another person's land, but can also be committed when objects are thrown across boundaries. Even a bullet, fired quite legally on property where the shooter has legal access but which lands beyond its boundary, can be defined as an act of trespass. In such a way a poacher cannot shoot from a public right of way onto adjoining land and then claim this as a defence.

Generally speaking, trespass is a civil matter and the police have no powers to act if no crime has been committed. However, some specific legislation creates criminal offences where private property may be occupied without the owner's consent, such as by unauthorised campers or travellers.[9]

Other legislation covers the offence of aggravated trespass, which occurs when someone does anything to obstruct or disrupt a lawful activity, or acts with the specific intention of intimidating another person to the point that they are deterred from carrying out a lawful activity.[10] In such cases, individuals effectively forfeit their right to be on the land and instead become trespassers, and the police are empowered to take action and in some circumstances have powers of arrest.

While any landowner, or their agent, is perfectly entitled to require that a trespasser leaves their land, they have no powers to demand names and addresses. Although common law permits the physical removal of those who refuse to leave, such an approach must involve no more force than is reasonable and necessary, and carries the risk of a counter-claim for assault.

Poaching

Without the consent of the owner or occupier or other lawful authority, it is an offence for anyone to enter any land with the intention of killing, taking or injuring any deer, or to attempt to do so. Likewise, it is an offence to search for any deer with such intent, as it is to remove the carcase of any deer. The only defence is that someone might be able to prove the belief that the owner's consent might have been given to such actions had they known of the circumstances. The offences are comprehensively covered by separate but similar legislation for England and Wales,[11] Northern Ireland[12] and Scotland.[13]

9 Trespass (Scotland) Act 1865 and Public Order Act 1986

10 Criminal Justice and Public Order Act 1994

11 Deer Act 1991 Section 1 (1)

12 Wildlife (Northern Ireland) Order 1985, Section 22

13 Deer (Scotland) Act 1996, Section 17

14 Control of Noise at Work Regulations 2005

In the eyes of the law, any 'authorised person', being the owner or occupier of the land or someone authorised by them (this includes people who have the legal right to kill or take deer on that land) may, if they have reasonable cause to suspect an offence, demand the name and address of the offender and order them to quit the land immediately. Failure to comply with this is an additional offence in its own right. However, as poachers frequently operate in gangs and are capable of offering violence, threats and intimidation, it is not usually advisable to follow this course.

Noise at work regulations

Anyone employing others needs to be aware of legislation controlling noise at work,[14] designed to protect the hearing of workers against unnecessary damage in the workplace. Appropriate actions need to be taken if noise levels exceed 85 decibels (dB) of sound, or provide information and training if levels exceed 80dB.

An unmoderated rifle produces over 160 decibels (dB) of sound, often considerably more depending on the rifle and calibre in use; adding a moderator can reduce this to as low as 130dB. For comparison, normal conversation is rated at around 60dB, a vacuum cleaner is about 70dB, and a chainsaw some 115dB. Many employers, such as the Forestry Commission and self-employed people bound by these regulations, have found that the provision of sound moderators has enabled them to comply.

Though the regulations do not apply to individuals exposed to noise from their non-work activities, or when they make an informed choice to go to noisy places

⬇ *A wide variety of sound moderators can be fitted to rifles to reduce their sound signature. The two models on the right are designed to sleeve back over the barrel to reduce the overall length.*

or from nuisance noise, it is only common sense that even those who are not acting in the capacity of employee should take appropriate action to protect their hearing when working in noisy environments. More information and guidance is of course available from the Health and Safety Executive.

Knives

While a knife may be regarded by the average user as no more than a working tool, it is important to recognise that the carriage and use of one may be controlled by legislation under which illegal possession may result in a heavy fine or even imprisonment.

In essence, it is illegal to carry a knife in a public place without good reason unless it has a folding blade with a cutting edge of 3 inches or less. 'Good reason' may include possessing it at work, or carrying it to or from work. While there are many types of knife that are banned completely (suck as flick knives, butterfly knives, sword sticks and a long list of others), these are unlikely to be used for deer management activities in any case.

➡ *Clockwise from left: fixed blade, locking (mechanism indicated by arrow) and folding knives. Only the last, having a cutting blade of less than 3 inches, can legally be carried in a public place without good reason.*

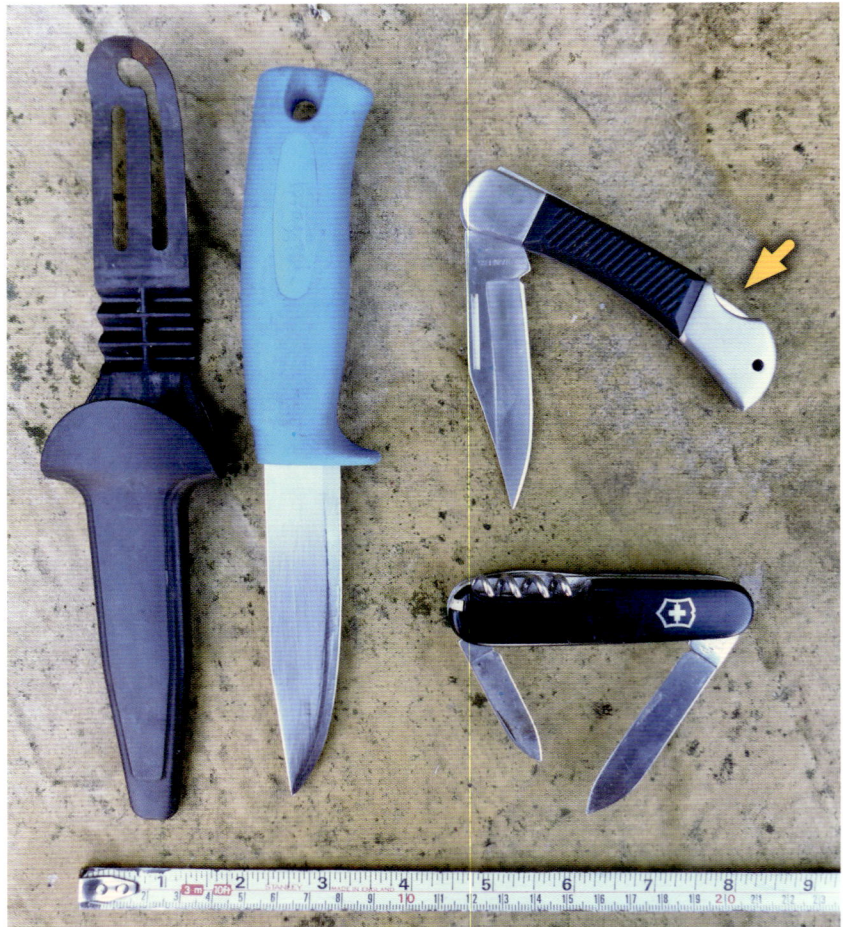

Lock knives (i.e. those which can be locked and refolded simply by pressing a button) are not classified as folding knives irrespective of blade length; this definition also includes multi-tool knives, whose attachments may also perform other functions, such as files, screwdrivers and can openers. It follows that these, and the fixed blade knives commonly used by deer managers, should be carried with care. It is recommended that they are best packed away in rucksacks with other working items rather than being carried on the person, to avoid misunderstanding or inadvertently having them with you in inappropriate places.

•••••••••

The points covered above are the main areas of which anyone involved in the management of deer needs to be aware. Further reading on the subject is recommended in Appendix I. At a working level you do not need to know the law inside out, but you do need to be properly aware of what it requires. As always, if in doubt seek advice.

*Finally, it cannot be stressed strongly enough that you (and those working with you) need to be appropriately insured **at all times**.*

Endnote

Prior to 1945, deer management as we know it today was not widely practised in much of the United Kingdom. Deer numbers were low in any case, kept in check by a combination of factors which included different approaches to land management and use, a lack of habitat and, where they did occur in any quantity, some very indiscriminate shooting. Deer were certainly not widely recognised as assets and their treatment in many places amounted to little more than that given to vermin. Far more may have been killed by shotguns and snares than ever fell to a rifle. Only really in Scotland was there a well-developed appreciation of deer and even then, that largely only applied to the red deer in a sporting sense. Elsewhere, the concept of deer as a valuable, renewable resource with their own place in the natural landscape – rather than as decorative assets confined to parks - was rare. The alien introductions of the late nineteenth century were still regarded as little more than curiosities, existing only in small numbers and confined to very limited parts of the country.

Things started to change largely after servicemen started to return from Europe at the end of the Second World War. They had experienced the high respect paid to deer as a sporting quarry in many countries, notably Austria and Germany, where management was more highly developed and formalised, and brought some of the practices they had seen back to this country with them. They, and others who thought like them, started to look at deer with newly opened eyes. Change did not occur overnight, however, and it was not until the Deer Act of 1963 that deer finally started to receive any degree of proper protection and humane treatment.

When I first became seriously involved with deer in the very early 1980s it would be fair to say that deer management practices, in the southern parts of Britain at least, were still evolving and the majority of professionals were to be found in Scotland or with specialist organisations like the Forestry Commission. Otherwise experienced, dedicated stalkers were few and far between, and training was only available if you were lucky enough to find a suitable mentor or belonged to one of the very few management groups that existed. Available literature in the English language was largely restricted to just a few handbooks but, thanks to the work of authorities such as Richard Prior and Andrew de Nahlik, an increasing amount of knowledge was starting to become more widely available. Outside the small but growing circle of deer aficionados, though, attitudes could still be less enlightened. A friend who attended a two-year forestry course at the time later told me that deer were covered in one brief, half-hour lecture, which essentially encouraged an approach largely based on shooting on sight.

How things have changed over only a few more decades: our attitude towards deer has improved radically as we have come to understand and appreciate them better. During that time roe and fallow numbers have grown considerably across the southern half of England, the muntjac no longer occupies a range restricted simply to the Midlands and East Anglia and is increasingly encountered in suburbia, while even the sika is starting to turn up in unexpected places. In Scotland, a drive towards landscape regeneration has forced a re-evaluation of attitudes towards red deer in particular. Only the diminutive and less robust Chinese water deer is still comparatively less numerous – but there are signs that even they are slowly increasing their range.

Fortunately, wider management practices have improved. The British Deer Society, promoting deer welfare while accepting a practical need for their numbers to be controlled, has been at the forefront of developing stalker training to ensure competence as well as enhancing best practice and humane treatment. The year 1982 saw the introduction of the Woodland Stalkers' Competence Certificate, which evolved via the National Stalkers' Competence Certificate into the internationally recognised Deer Stalking Certificate that we know today. By 2018 almost 26,000 stalkers had achieved the DSC1 with more than 5,400 successfully meeting the standards for DSC2. In the meantime, venison, long considered a luxury food for the privileged, is finally widely recognised as a readily-available, healthy, nutritious and delicious source of protein, and has surged in popularity among the greater public.

By the last decade of the twentieth century more official interest was also being taken in what was finally being seen as an emerging issue. In Scotland the problem of damage caused by marauding red deer had long been recognised, leading to the creation in 1959 of the Red Deer Commission with special powers to protect both forestry and agriculture. This in turn became the Deer Commission for Scotland with a much wider responsibility for the conservation, control and sustainable management of all species of wild deer in Scotland. Its functions were transferred to Scottish Natural Heritage in 2010. England and Wales saw the formation of the Deer Initiative in 1995 as a broad partnership of statutory, voluntary and private organisations with a shared vision for the sustainable management of wild deer.

At the same time the public has become increasingly aware of the wider natural world, thanks in part to a highly developed media and an increasingly sophisticated and vocal animal welfare lobby. In many cases this has encouraged some unbalanced views which range between the absolute protectionist who thinks that it is wrong to kill such beautiful animals for whatever reason, to those who demonise deer because of their effects on the habitats of other creatures. Clearly there has to be a middle way.

As we have seen, although it is possible to mitigate against deer issues in non-lethal ways, the fact remains that some problems may only be postponed or transferred elsewhere. Like it or not, we have to accept that deer have to be controlled to keep their numbers in check. If there are no apex predators within an ecosystem to do this, the job must fall to man and until science is able to offer a viable alternative the rifle must, of necessity, remain at the forefront of the available tools for the task.

The modern deer manager is usually a mixture of stalker, ecologist, countryman and deer lover who recognises that understanding and respect are the cornerstones of any intelligent approach to deer and the issues that they might raise. Deer remain a valuable part of our countryside which would be a poorer place without them. They are a key feature of the rich biodiversity of the United Kingdom's landscape.

Long may they remain so.

Appendix I Further Reading

Of the multitude of books available on the subject of deer and their management, I detail below some personal recommendations. Some may now be out of print but a visit to the websites of Coch-y-Bonddu Books at www.anglebooks.com or Amazon at www.amazon.co.uk will source both new and used books.

General books on deer

Alcock, Ian *Deer – A Personal View* (Swan Hill Press 1996)
A beautifully written appreciation of deer by a respected observer.

Chapman, Donald and Norma *Fallow Deer* (Coch-y-Bonddu Books 1997)
A detailed and learned guide to their history, distribution and biology by two highly respected authors.

Chapman, Norma *Deer* (Whittet Books 1991)
A comprehensive guide to the origins and habits of British deer.

Geist, Valerius *Deer of the World* (Swan Hill Press 1999)
A detailed and scholarly examination of deer evolution and behaviour.

Prior, Richard *Deer Watch* (Swan Hill Press 2007)
An overview of the origins, history and ecology of British deer.

Prior, Richard *The Roe Deer* (Swan Hill Press 1995)
A major and comprehensive appreciation of the natural history and all aspects of dealing with roe, by an acknowledged expert on the species.

Putman, Rory *The Natural History of Deer* (Christopher Helm 1988)
A detailed overview of deer origins, behaviour and natural history by one of our foremost experts. Out of print but well worth searching for.

Smith-Jones, Charles *Muntjac: Managing an Alien Species* (Coch-y-Bonddu Books 2004)
A guide to the natural history, stalking and management of this elusive and potentially invasive deer.

Whitehead, G.K. *The Whitehead Encyclopedia of Deer* (Swan Hill Press 1993)
An extensive and accessible work covering hunting, biology, history and a host of other aspects relating to deer, this is probably the ultimate in reference material for any deer *aficionado*. For a long time unavailable, but reprinted in 2008 by popular demand.

Currently there are no books devoted specifically to sika or Chinese water deer in the UK; hopefully in the future someone with the suitable knowledge and experience of these species will remedy this.

Management

Griffiths, Dominic *Deer Management in the UK* (Quiller Publishing 2011)
A guide to managing deer to their best potential from a highly knowledgeable and authoritative professional, with the emphasis on roe.

McShea, J.M. et al *The Science of Overabundance* (Smithsonian Institution 1997) An analysis of deer ecology and population management by a number of leading ungulate biologists. Although demanding reading and focusing on white-tailed deer in the USA, this collection of papers has great relevance to deer management in the UK and is essential reading for anyone with an academic interest in the subject.

de Nahlik, A.J. *Deer Management* (David & Charles 1974)
A classic work on the subject.

Putman, Rory *The Deer Manager's Companion* (Swan Hill Press 2003)
A guide to the management of deer in the wild and in parks.

Stalking

Downing, Graham *The Deer Stalking Handbook* (Quiller Publishing 2004)
An excellent introduction to deer stalking.

Deer stalker training

British Deer Society *Training Manual for Deer Stalkers* (BDS, regularly updated) Essential reading for anyone attempting the Deer Stalking Certificate 1, this handbook is also a superb source of reference for any deer stalker.

Tracking dogs

Sondergaard, Niels *Working with dogs for deer* (Jagtforlaget 2006)
A definitive guide from a leading Danish expert on training and working with deer-tracking dogs.

Jeanneney, John *Tracking dogs for finding wounded deer* (Teckel Time 2006) Another well respected work, covering all aspects of training with a special focus on the teckel or working dachshund.

Miscellaneous

Coles, Charles *Gardens and Deer* (Swan Hill Press 1997)
A clearly written and straightforward guide to deer damage limitation in gardens.

Parkes, Charlie and Thornley, John *Deer: Law and Liabilities* (2nd Edition, Quiller Publishing 2008)
An essential guide to the law relating to deer and deer stalking. Every person with deer management interests should own a copy.

Prior, Richard *Trees & Deer* (Swan Hill Press 1994)
A comprehensive guide to tackling the problems that deer pose in forest, field and garden.

Online resources

Deer Initiative Best Practice Guidance
A full and comprehensive set of guides on subjects ranging from deer ecology, culling, management planning, legislation and a wide range of other subjects. Available at www.thedeerinitiative.co.uk/best_practice

Best Practice Guide
As above, but issued by a partnership of interested partners for Scotland. Available at www.bestpracticeguides.org.uk

Ultimate Deer Data
Online training available only from the British Deer Society. An interactive training package enabling candidates to acquire and test their knowledge in preparation for taking the Deer Stalking Certificate 1. It details the requirements for all five assessments and provides comprehensive information on a massive range of deer-related subjects. Whilst UDD is primarily aimed at people wanting to achieve DSC1, due to the significant volume of and quality of its content it is of equal value as an important educational resource to anyone worldwide needing to gain knowledge of British deer and related management. For more details visit www.bds.org.uk

Appendix II ## Useful Addresses

Advice

The British Deer Society

A charity established in 1963 that works to ensure the six wild species of deer can exist in the UK today and also helps to secure their future. Among its objectives are research, education and deer manager training.

The British Deer Society, The Walled Garden, Burgate Manor, Fordingbridge, Hampshire SP6 1EF
Website: www.bds.org.uk
Email: h.q@bds.org.uk
Telephone: 01425 655434

The Deer Initiative

Established in 1995 to promote the sustainable management of wild deer in England, and was extended to Wales in 1999. A broad partnership of statutory, voluntary and private organisations with a shared vision for the sustainable management of wild deer.

The Deer Initiative, The Carriage House, Brynkinalt Business Centre, Chirk, Wrexham LL14 5NS
Website: www.thedeerinitiative.co.uk
Email: through website
Telephone: 01691 770888

Game & Wildlife Conservation Trust

A UK charity conducting conservation science to enhance the British countryside for public benefit. Established for over 80 years to research and develop game and wildlife management techniques; provides training and advice on how best to improve the biodiversity of the countryside. Formerly the Game Conservancy Trust.

The Game & Wildlife Conservation Trust, Burgate Manor, Fordingbridge, Hampshire SP6 1EF
Website: www.gwct.org.uk
Email: info@gwct.org.uk
Telephone: 01425 652381

Statutory bodies

The following are statutory bodies which (with the exception of APHA) are the first point of contact regarding deer legislation, and who are responsible for the provision of licences for night shooting etc. Further information is available on their websites.

Natural England

The government's adviser for the natural environment in England, helping to protect England's nature and landscapes for people to enjoy and for the services they provide. Natural England is an executive non-departmental public body, sponsored by the Department for Environment, Food & Rural Affairs (Defra).

Natural England, County Hall, Spetchley Road, Worcester WR5 2NP
Website: www.gov.uk/government/organisations/natural-england
Email: enquiries@naturalengland.org.uk
Telephone: 0300 060 3900

Scottish Natural Heritage

Funded by the Scottish government's Environment and Forestry Directorate, Scottish Natural Heritage promotes, cares for and improves Scotland's natural heritage.

Scottish Natural Heritage, Great Glen House, Leachkin Road, Inverness IV3 8NW
Website: www.nature.scot
Email: enquiries@snh.gov.uk
Telephone: 01463 725 000

Department of Agriculture, Environment and Rural Affairs Northern Ireland

Responsible for the law applying to deer in Northern Ireland.

Department of Agriculture, Environment and Rural Affairs, Wildlife Team, 2nd Floor, Klondyke Building, Cromac Avenue, Gasworks Business Park, Lower Ormeau Road, Belfast BT7 2JA
Website: www.daera-ni.gov.uk
Telephone: 028 9056 9558

Natural Resources Wales
Works on behalf of the Welsh government to ensure that the natural resources of Wales are sustainably maintained, enhanced and used.
Natural Resources Wales, Customer Care Centre, Ty Cambria, 29 Newport Rd, Cardiff CF24 0TP
Email: enquiries@naturalresourceswales.gov.uk
Website: www.naturalresources.wales
Telephone: 0300 065 3000

Animal & Plant Health Agency
An executive agency, sponsored by Defra, the Welsh government and the Scottish government, that works to safeguard animal and plant health for the benefit of people, the environment and the economy. Contact details for regional offices are available on the website.
Animal & Plant Health Agency, Woodham Lane, Addlestone, Surrey KT15 3NB
Website: www.gov.uk/government/organisations/animal-and-plant-health-agency

Tracking dogs

UK Deer Tracking and Recovery
A non-profit-making organisation which maintains a register of dogs and handlers which have passed examination on a 20-hour cold trail. Running regular training events, it aims to place as many dog and handler teams as possible at the disposal of estate managers, deer managers and stalkers in as many parts of the UK as possible. Further details can be found at www.ukdeertrackandrecovery.co.uk

Bavarian Mountain Hound Society of Great Britain
Offers training advice and courses for the training of tracking dogs, of all breeds, with representation on a regional basis. Further details can be found at www.bmhs.org.uk

Deer management training providers

A number of organisations and small businesses offer deer management training and qualifications. Some are better than others but the following are especially recommended:

Deer Management Qualifications Ltd
Administer a number of Assessment Centres for **Deer Stalking Certificates 1 and 2**. These deliver the relevant training and assessment. To view the full list, visit their website at www.dmq.org.uk

The British Deer Society
In addition to being an Assessment Centre for **DSC1 and 2**, the BDS is the sole provider of the **Deer Management Course** which is tailored for stalkers, landowners and managers – the decision-makers dealing with deer management issues within a variety of habitats. It is also suitable for individuals who might during the course of their work need to advise others on deer management matters and for those simply wishing to extend their general deer management knowledge. The course is usually of three days' duration and is certified by the nationally recognised awards body Lantra. For more details visit www.bds.org.uk

DRAFT CULL PLAN

Location: .... Species: ....

PART A. Census as at (date): .... Method: ....

Step		MALE	FEMALE
1	**Numbers seen**		
2	**Add % not seen _____%** (Guide 10-30%) (Notes 1, 2 & 3)		
3	**TOTAL** (Steps 1 & 2))		

PART B. **Theoretical Population Increase**

Step		MALE	FEMALE	KIDS (M&F)
4	**Number of females that breed _____%** (Guide 60%) (Step 3 x %)			
5	**Multiply breeding females (Step 4)** **by fecundity rate used: _____** (see Note 4)			
6	**Divide result of Step 5 into M/F** (Guide 1:1)			
7	**Other forecast GAINS** (immigration etc)			
8	**Forecast LOSSES** (winter/early mortality, poaching, RTA, emigration etc)	Minus _____	Minus _____	
9	**Gross population by 1 July** (Step 3 + Step 6 & 7 minus Step 8)			

PART C. **Draft Cull Plan**

Step		MALE	FEMALE	
10	**Carrying capacity**			See Note 5
11	**Planned Cull** (Subtract Step 10 from Step 9)			
12	**Breakdown of Cull:** a. Young _____ % (Guide 60%) b. Mature _____ % (Guide 20%) c. Old _____ % (Guide 20%)			See Note 6

Appendix III Draft Cull Plan

NOTES:

1. Any fractions calculated (Steps 2, 4, 5 & 6) should be rounded up.

2. Guidance on % figures are given throughout in brackets. Enter the actual figure used in the space provided.

3. The chosen percentages for animals not seen (Step 2), breeding percentage of female population (Step 4), and numbers for Gains & Losses (Steps 7 & 8) will all be dependent on local circumstances.

4. Fecundity rates will vary between deer species; suggested figures are below but may need to be adjusted further taking local factors into account:

    a. Muntjac   1.7
    b. CWD   1.8
    c. Roe   1.6
    d. Fallow   0.9
    e. Sika   0.9
    f. Red   0.8

5. Carrying capacity (Step 10) is determined according to local circumstances.

6. The recommended breakdown of the proposed cull into age categories is as follows:

	Large deer species	**Roe & muntjac**	**CWD**
Young	Not yet 3 yrs	Not yet 3 yrs	Not yet 1 yr
Mature	Not yet 9 yrs	Not yet 7 yrs	Not yet 5 yrs
Old	Over 9 yrs	Over 7 yrs	Over 5 yrs

Where an imbalanced deer population structure exists, it may be necessary to alter the cull by age categories away from the suggested % shown in the table.

Special care should be taken, and allowances made, when planning culls for species which move seasonally (especially fallow, and red in some areas).

Culling by age categories is of lesser importance for female deer, which are usually culled as seen within planned targets (but with due consideration to issues such as orphaning dependent young etc).

Special note: the Planned Cull must not be seen as a fixed figure. It should be considered as a flexible target which can be adjusted, as the culling year progresses, according to any variations, seasonal or otherwise, that occur within the dynamics of the deer population.

Appendix IV **Sample Risk Assessment**

Location: *Cold Comfort Farm*   **Date of Risk Assessment** *31 March 2019*

Hazard	At risk	Controls
Gunshot injury	Stalkers, other estate staff, public	Correct load/unload procedures Vigilance before shooting Only shoot with known backstop Ricochet hazard awareness Maximum use of high seats Warning signs on footpaths when culling in progress
Fall from high seat	Stalkers, other users, general public	Check before use Wire all rungs Non-slip surfaces Warning signs Site seats away from public areas Regular inspection
Medical emergency	Stalkers	Booking out/in procedure Individual stalkers notify any significant medical issues All carry small first aid kit and mobile phone
Transport on site	Stalkers and others	Injury if struck by vehicle All operators to hold appropriate licence ATVs to be operated only by trained persons All vehicles maintained and inspected according to manufacturers' instructions Seatbelts (if fitted) to be worn at all times
Zoonoses	Stalkers	Risk of transmission of diseases from animals Protective clothing as appropriate, especially rubber gloves for all gralloching Hot and cold water, soap and paper towels available in farm rest room All to wash hands after contact with animals and especially before eating, drinking or smoking
Contamination	Stalkers, public	Carcase preparation in accordance with best practice Clean equipment Carcase extraction by hygienic means
Lifting weights	Stalkers	Correct lifting techniques Use of winches for larger carcases Seek help if necessary
Knife injuries	Stalkers	Cut away from self Awareness of those around you Sharp and clean knives
Third party injury	Stalkers, other estate staff, public	All to follow operating procedures Stalking insurance

Further action required?	Actions:		
	Responsible	Due date	Date completed
All unaccompanied stalkers to hold at least DSC1 All guests to be accompanied	All stalkers	Ongoing	Ongoing
Annual maintenance and high seat register	Head stalker	Annually each March	12 Mar 2019
Consider first aid course	All stalkers	Ongoing	N/A
Confirm holders of ATV training certificates	All	Ongoing	30 Mar 2019
	All	Ongoing	N/A
	All	Ongoing	N/A
	All	Ongoing	N/A
	All	Ongoing	N/A
Check stalking insurance for all stalkers	Head stalker	Annually each March	12 Mar 2019

Appendix V **Sample Deer Management Report**

CONFIDENTIAL

Cold Comfort Estate
Deer Management Group

The Bothy
Anytown
Blankshire EF12 3GH
Tel (01234) 891567
jsmith101@internet.com

Mr AB Factor
The Estate Manager
Cold Comfort Hall
Blankton
Blankshire WX56 7YZ

20 March 2019

**ANNUAL REPORT FOR COLD COMFORT ESTATE
2018/2019 (APRIL TO MARCH)**

General

1. It has been a successful year and all cull objectives have been met. The roe doe cull was, as in the previous year, a particular challenge with few animals showing before the New Year due to a combination of winter inappetence, mild weather and abundant forage in the woods. As a result animals showed an understandable reluctance to leave cover and much movement until then appears to have been nocturnal.

2. A map of the estate is attached at Annexe A.

Environment

3. The management area on Cold Comfort Estate covers an area of approximately 1,900 acres, consisting of woodland and mixed arable farmland. The habitat available to deer also includes open set-aside fields and scrub. In addition, the Shoot maintains some 15 acres of cover crops, mainly maize with some double turnips, sunflower and millet, which are used by the deer for browse and cover.

4. The annual scrub and ride clearance has taken place during the past year. There are no major areas of newly planted trees, and deer damage is not considered to be a concern at the present time.

5. Suspicions of poaching activities remain unproven. Unidentified vehicle tracks have been noted on several occasions and a roe doe culled in December 2018 was found to be carrying SSG-sized lead shot. Hard evidence as to the perpetrators is not easy to come by, and all remain vigilant.

CONFIDENTIAL

CONFIDENTIAL

Deer Population

6. Roe Deer

    a. The roe population at the beginning of the season was estimated at some 83 animals with the holding capacity set at 70. After taking forecast increases into account a cull of 18 bucks and 15 does, increased to 22 bucks and 19 does after a re-assessment of numbers in June 2018, was set and achieved. Over the past 12 months there has been no reported deer mortality outside the set cull.

    b. The overall population is healthy, maintaining high body weights even among yearling animals during the winter. Despite high area usage, stress does not appear to be an issue at present as there is usually a quiet corner available for deer to lie up in. Ticks and keds are generally seen in low numbers, with little lice infestation apparent. Apart from one instance of lungworm found in an old buck, there have been no cases of internal parasites.

    d. The ratio of bucks to does has levelled out. However, an imbalance in the age structure among bucks, as identified in last year's report, remains. Of the 22 bucks culled in 2018, 6 were in the 'old' category. Attention will continue to be paid to this during the coming season. The estate contains at least three bucks of notably high quality (both in body weight and antler formation) which will be carefully preserved.

    e. Historical fecundity figures used for forecasting are considered to be accurate and have been held at 1.6. The figure for deer unseen has been set at 15% for census purposes.

    f. The 2019-20 roe cull plan is at Annexe B. It is set against an estimated population, based on a rolling census and a thermal imaging count conducted in early March 2019, of 89 roe.

7. Muntjac Deer

    a. Sightings have increased over the past six months and 9 muntjac have been culled since April 2018, compared to a maximum of 2 per year over the past few years. This suggests an overall increase in the estate muntjac population, which was previously estimated at no more than 10 – 15 animals. Of the culled animals, 6 were bucks and other sightings tend to reflect this gender imbalance. It is believed that the muntjac population in Blankshire has grown, with the subsequent expulsion of

CONFIDENTIAL

predominantly young bucks seeking new territory. Although it is not felt that muntjac have any significant impact on the environment at present, the potential increase in numbers is a cause for mild concern and developments will be monitored carefully.

c. Muntjac will continue to be culled on sight in line with British Deer Society guidelines.

8. Fallow Deer

a. There is no established fallow population on the estate. A melanistic fallow doe was sighted in June 2018 and there have been several sightings of a similar fallow doe since then and it is assumed that these have been of the same animal. The nearest fallow population is five miles away in the area of Stately Park.

b. In line with estate policy, fallow deer will not be encouraged and will be culled as seen according to season, although any mature bucks will be left as agreed with the Blankshire Deer Management Group (BDMG).

9. A consolidated list of cull reports for 2018/19 is at Annexe C.

Liaison

10. Liaison continues to be maintained with all parties interested in land usage, including farm staff and the Cold Comfort Shoot. All are kept abreast of the Group's activities. Meetings of the BDMG are attended by the head deer manager, enabling full liaison with all neighbouring deer management interests.

11. All access for stalking and associated purposes is confirmed with estate staff prior to the event. Signs are placed as appropriate to provide visual warning to other land users that stalking is taking place, and all carry mobile telephones and contact details for the Estate Manager and tenant farmer at all times.

12. A Risk Assessment has been conducted in consultation with the Estate Manager and is attached at Annexe D.

13. Particular emphasis continues to be placed on liaison with the Shoot. Care is taken to ensure that stalkers on the ground on game shooting days exit via the Shoot Hut to inform their staff that the former are off the ground. With considerate liaison on both sides, clashes of interest will continue to be avoided.

CONFIDENTIAL

Training and Infrastructure

14. Rifles are check-zeroed regularly at times and locations agreed with the Estate Manager. All stalkers have now achieved the Deer Stalking Certificate 2, and Mr J Smith attended and passed the BDS Deer Manager's Course in August 2018.

15. Two new wooden high seats have been erected to cover places where the tenant farmer has complained of increased deer damage. All seats have been inspected and a Certificate of Serviceability is at Annexe E.

Conclusion

16. The roe buck and doe culls for 2018-2019 have been achieved. There appears to have been an increase in muntjac numbers and there is evidence of a very occasional transient fallow presence. Overall, the Cold Comfort Estate deer population is considered to be stable and healthy.

(Signature)

Annexes:

A. Estate Map (including high seat locations)
B. Census and Cull Plan
    (Note: see Chapter 7 and Appendix III *for a suggested format)*
C. Consolidated List of Cull Reports
    (Note: this is best presented as tables arranged by species and sex. Details contained can be as simple as the date and location of individual culls, but other useful information might include age, weight, heath, kidney fat %, parasites, antler formation and length, range of shot, wound site and reaction, dealer price received etc)
D. Risk Assessment
    (Note: see Chapter 4 and the suggested format at Appendix IV*)*
E. High Seat Grid References and Certificate of Serviceability
    (Note: this should include a simple description of the type of seat, a description of its location and a grid reference. Estimates of each seat's estimated life and any other remarks are also useful. The list should conclude with a signed statement to the effect that 'all of the above high seats have been inspected, are marked with the necessary warning sign and are in a serviceable condition unless stated otherwise'.)

CONFIDENTIAL

Appendix VI Simple Letter of Permission

To whom it may concern

PERMISSION TO SHOOT ON COLD COMFORT ESTATE

Reference: Cold Comfort Estate boundaries map (attached)

This letter confirms that Mr **John SMITH** of The Bothy, Anytown, Blankshire EF12 3GH is authorised to control deer and vermin within the boundaries of Cold Comfort Estate (the Estate) and may use firearms for this purpose subject to the following conditions:

• **Deer Management.** While Mr Smith is granted primary responsibility for deer management within the Estate, all sporting rights remain the property of the Estate owners.

• **Game shooting.** Guests of the Estate may, from time to time, require access for the shooting of game and other purposes. Mr Smith will be given advance notice of any such access.

• **Notification.** Prior notification is to be given to the Farm Manager (FM) or his nominated representative by telephone before Mr Smith enters the boundaries of the Estate.

• **Timings.** Access to the ground will normally be between one hour before until three hours after sunrise, and three hours before until one hour after sunset. Any variations to these timings should be agreed with the FM beforehand.

• **Access.** Normal access to the Estate is to be via the South Lodge entrance at GR SU123456. Other access points are only to be used after prior arrangement with the FM.

• **Restricted areas.** The immediate area of Cold Comfort Hall as marked on the attached map is not to be accessed without prior agreement.

• **Vehicles.** Vehicles may only be driven on the estate roads and established gravelled tracks. If vehicles are to be taken off-road the agreement of the FM is to be sought first.

• **Firearms.** All firearms used are to be appropriately licensed, suitable to the purpose they are used for and moderated where possible to minimise noise and associated disturbance to the surrounding area.

• **Infrastructure.**

a. **Positioning.** The positioning of high seats or any other infrastructure items is to be discussed with the FM before it takes place.

b. **Associated Activities.** Associated activities, such as the pruning or felling of undergrowth to permit sight lines, are likewise to be agreed.

c. **Inspection.** All infrastructure is to be formally inspected for serviceability at intervals not exceeding 12 months.

• **Risk Assessment.** Mr Smith is responsible for ensuring that appropriate risk assessments are completed before any activities take place.

• **Assistants and guests.** The agreement of the FM is to be sought before any assistants may accompany Mr Smith within the Estate. Mr Smith will be responsible for ensuring that they are suitably supervised.

• **Legal/Best Practice.** All activities relating to this authority are to be conducted according to the law and prevailing best practice/codes of conduct.

• **Insurance.** Mr Smith is to ensure that he possesses appropriate levels of insurance relevant to his activities and those of any assistants.

• **Carcases.** Proceeds from the sale of carcases is to be split 50/50 between Mr Smith and the Estate.

• **Cull Planning.** Cull targets for the coming year are to be agreed between Mr Smith and the Estate every March.

• **Records.** A record of all culls (by species, sex, age and location) is to be maintained.

• **Reporting.**

  a. **Culling.** A report of all deer and vermin culled is to be submitted to the FM every six months, or at any other time if requested, along with any relevant proceeds from sales.

  b. **Damage etc.** Any instances of damage to property, unusual events or suspicious activities are to be reported to the FM without undue delay.

This permission may be terminated at the discretion of Cold Comfort Estate at any time and without prior notice.

(Signature)

AB Factor, Estate Manager
Cold Comfort Hall, Blankton, Blankshire WX56 7YZ
01234 567890
coldcomfort@anyinternet.com

Appendix VII Suggested Briefing Notes for Group Activities

The following are suggested subjects that should be covered when briefing groups for activities such as censuses, ground clearance or building projects, or collaborative deer culls. Not all may apply and there may be other headings that you wish to add according to the circumstances.

They are laid out in a logical sequence, and in many cases a verbal briefing would be enhanced by a written summary, photocopied map (annotated as appropriate with movement lines, shooting positions, high seat locations etc) and a copy of the risk assessment that you have completed for the activity. This should be provided to each person taking part. The inclusion of written permission to all involved in shooting is also helpful in case anyone challenges them.

Ground

A general description of the ground to be covered (maps are particularly useful here), with special emphasis on:

❯ Landmarks
❯ Potentially dangerous areas such as roads or watercourses, and their most suitable crossing points
❯ Sensitive areas such as conservation sites or livestock penning, or other places to be avoided
❯ Public footpaths and other access
❯ Other land users and their forecast activities
❯ Routes to and from meeting areas

Aim

A simple statement of what you intend to achieve.

Conduct

A full description of how the activity will be carried out. This should include:

❯ An overall outline
❯ Detailed description of activities, broken down into stages for simplicity if necessary
❯ Responsibilities, including who will be where if split up
❯ Identification of names/codes for shooting or observation positions for use in communications
❯ Timings (start, finish, meals etc)
❯ Transport arrangements to and from fixed positions (if appropriate)
❯ Use of protective clothing, stressing the need to wear appropriate personal protective equipment (e.g. hard hats, working gloves, steel-toe-capped boots or ear defenders as appropriate). High-visibility jackets are especially useful for walking personnel on collaborative culls or moved census counts
❯ Control of dogs
❯ Any special procedures (e.g. for climbing into high seats)
❯ Emergency procedures (including signals and meeting points)
❯ A run through of the Risk Assessment for the activity
❯ Rendezvous arrangements on conclusion of the activity

If firearms are involved, special emphasis should be placed on:

◗ If culling, what species, sexes and age classes of deer may be shot, and what must not be taken
◗ When rifles may be loaded/must be unloaded
◗ When shooting may commence, and the time of the last shot
◗ Emergency 'cease fire' signal
◗ Maximum shooting ranges
◗ Suitable backstops
◗ Ricochet hazards
◗ Stationary targets only
◗ When high seats etc may be vacated (it is recommended that whenever collaborative efforts involve shooting, strict emphasis should be given to the need to remain in high seats at all times)
◗ Recording the locations of shot deer
◗ Actions in the event of wounded deer (including not attempting to follow up while shooting is still taking place, location of dog handlers, not disturbing scent trails etc)
◗ Arrangements for collecting carcases
◗ Arrangements for processing carcases

Note: under normal circumstances it is advisable that firearms should not be carried by anyone during census or other non-culling activities.

Control and communication

◗ Who has overall responsibility
◗ Any nominated deputies and their functions
◗ Radios – who hold them, simple call-signs if necessary ('Zero' for control and 'High Seat One', 'High Seat Two' etc are suggested)

◗ Radio frequencies in use, alternatives and the signal to change frequency if necessary
◗ Mobile telephone numbers if necessary
◗ Alternative emergency signals if radios not working or no mobile telephone signal
◗ Any other people who may be encountered (e.g. landowner, farm staff etc)
◗ How to engage with the public if approached, and who to refer them to
◗ Time and location for a debriefing at the end of the activity (if required)

Administration

◗ Any special equipment requirements (torches, binoculars, survival equipment if in remote/ harsh locations etc)
◗ Location of any central point for storing equipment, fuel etc
◗ First aid (including location of first aid kits and identification of trained personnel)
◗ Procedure for contacting emergency services, including a description and/or grid reference of the rendezvous for any ambulances or other vehicles
◗ Identification of a landing area for air ambulance if appropriate
◗ Location of closest hospital with A&E provision
◗ Vehicles and parking arrangements
◗ Meal arrangements
◗ For those involved in shooting, a physical check of relevant FACs and shooting insurance is advisable

Always finish with an opportunity for questions.

Index

Note: Page numbers in italics refer to illustrations, with minimal relevant text. There are many illustrations on other pages too.

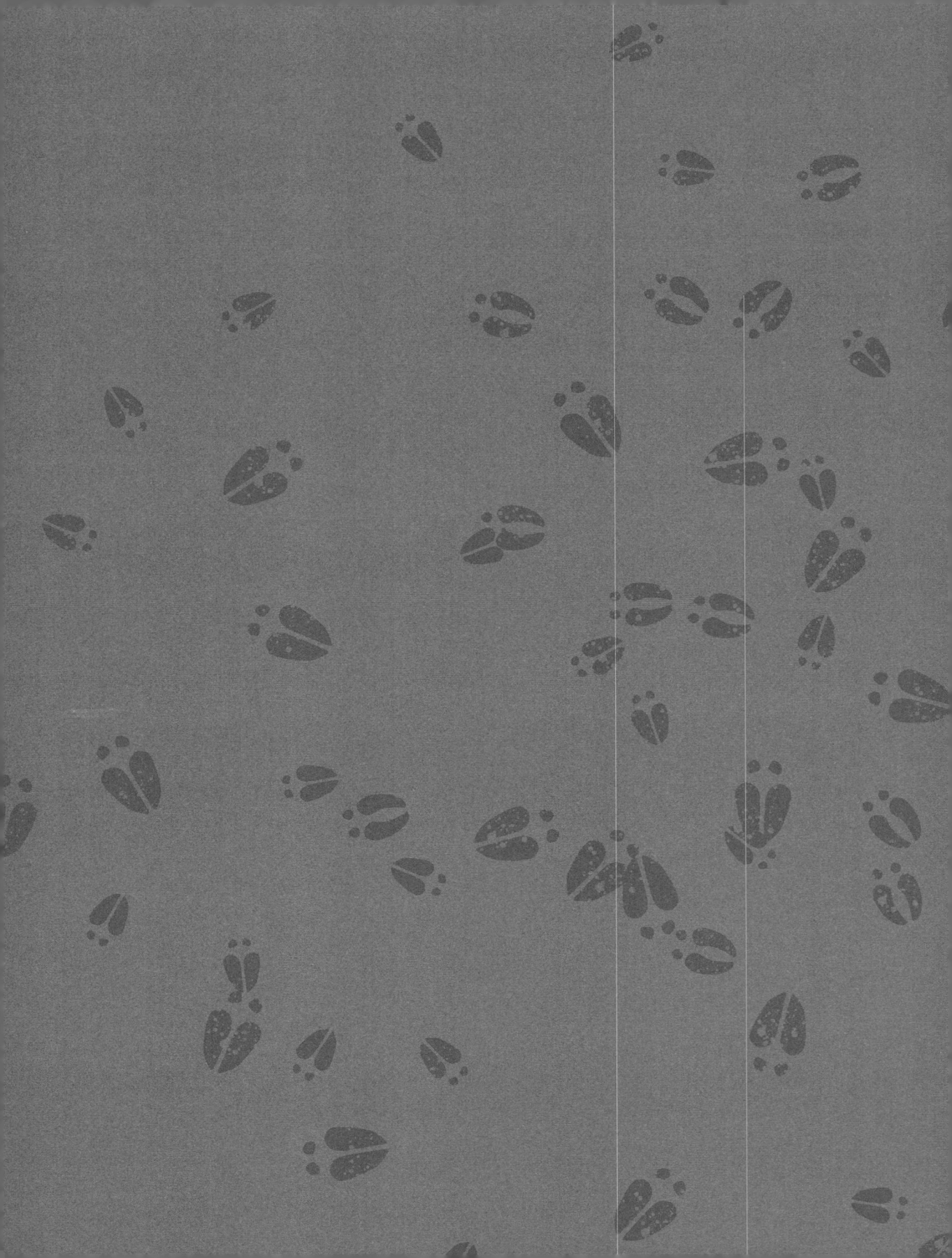